The Sexual Evolution

Also by Nathan H. Lents

Human Errors: A Panorama of Our Glitches, from Pointless Bones to Broken Genes

Not So Different: Finding Human Nature in Animals

The

SEXUAL
EVOLUTION

How 500 Million Years of Sex, Gender, and
Mating Shape Modern Relationships

Nathan H. Lents

MARINER BOOKS

New York   Boston

HarperCollins books may be purchased for educational, business, or sales promotional use. For information, please email the Special Markets Department at SPsales@harpercollins.com.

FIRST EDITION

Designed by Renata DiBiase
Illustrations by Don Ganley

Library of Congress Cataloging-in-Publication Data has been applied for.

ISBN 978-0-06-337544-4

24 25 26 27 28 LBC 5 4 3 2 1

I dedicate this to every queer person who has been told they are unnatural.
Not only are you natural, you are flawless and beautiful, exactly as you are.

There is grandeur in this view of life …
from so simple a beginning,
endless forms most beautiful and most wonderful
have been, and are being, evolved.

—Charles Darwin, *On the Origin of Species*, 1859

Contents

The Sexual Evolution

Introduction
The State of Affairs

(if you'll pardon the pun)

One day in my high school biology class, the teacher announced that there would be a closed-book *assessment* at the end of the week covering the content of the chapter we'd been discussing. This was the early 1990s, so my classmates and I had not heard that term before. Dr. Parrish explained the kinds of questions that would appear on the assessment and gave us tips for focusing our studies.

Okay, but is this a test? we asked. Dr. Parrish responded that the assessment would be worth 20 points and reminded us that the quarter's work comprised about 200 points in total, not counting extra credit opportunities.

So is this a quiz then? Her patience beginning to fade, Dr. Parrish explained that it was worth about 10 percent of the total points of the term, so we should study accordingly.

Why won't you just tell us if this is a test or a quiz???

I just told you how much it's worth! What difference does it make what you call it?!

Of course, it makes all the difference in the world. Humans have a strong instinct to name, categorize, and organize everything in our lives. This helps us make sense of the world around us, both the natural world and the world that we create. But these labels are artificial. They are *constructs*, which is not to imply that labels and categories are meaningless, just that they don't exist as real entities until we create them. There is no inherent difference between a test and a quiz. Those are just different words used to indicate the relative size and value of an assessment. By giving us the point value, Dr. Parrish had rendered the labels superfluous, but this did not satisfy us. We needed to know what we were studying for, a test or a quiz. At least in our minds, the two were different.

In the field of biology, we have a system—a whole discipline of study—that is all to do with the naming and classification of organisms. But these classifications, called *taxonomy*, are purely human creations and serve only our need to understand and communicate about the organisms that we study. We tend to think of a "species" as a real thing. This is a wolf, *Canis lupus*, and that is a dog, *Canis familiaris*. They are related, but they are different. You are a modern human, *Homo sapiens*, but the skullcap found in the Neander Valley on that fateful day in 1856 belongs to a different species we call *Homo neanderthalensis*. We are related, but we are different. A *species* includes all the members of one type of thing that are not members of another type of thing.* Simple. Neat.

Except . . . we know that dogs were domesticated (or self-domesticated) from wild wolves. Long ago, some wolves hanging around groups of humans underwent rounds of selection for thousands of years until the population had distinct enough features that they weren't really wolves anymore. At what point did they stop being wolves and start being dogs? Was there a time at which the last generation of wolves gave birth to the first generation of dogs?

Similarly, modern humans and Neanderthals are both descended from a common ancestor species that most anthropologists call *Homo heidelbergensis* (although some believe it was *H. antecessor*). While *our* ancestors remained in Africa and continued to evolve there, the ancestors of Neanderthals evolved mostly in the Near East, Central Asia, and Europe. At some point, the two simultaneously evolving populations became so different from each other that we recognize them as different things and we assign them different species names. But at what point do we draw that line? If we could recover an unbroken line of ancestors from *H. heidelbergensis* to us, would we be able to pinpoint the very first modern human? The answer is no, we couldn't. So what good are the categories if the borders are so blurry that we can't draw lines between them?

---

* This is, obviously, not the scientific definition of *species*. But so difficult is this concept to define that there are several competing definitions within the field of biology.

The problem here is that the differences among species are what we call *continuous*, especially if you include ancestral populations. There really is no possible way to draw a hard line, but this does not mean that the differences themselves are not real and that the categories are meaningless. Consider this parallel: developmental age. At some point in the past, you were a child, and you possessed the features that children have: small stature, physical and emotional immaturity, and so on. Currently, you have the features that adults have. At what point did you become an adult? Was there a day on which you went to bed a child and awoke as an adult? Of course not. It was a continual process with fuzzy borders, but that does not mean that adults and children are the same. We all know that adults and children are different and must be treated as such.

To make sense of the variation we see in sex, gender, and sexuality, we are going to have to become comfortable with the concept of continuous categories, because that's how life really works. Labels and categories are things that we construct so that we have a way to communicate about differences, but the realities of life weave in and out of those categories with little regard for our feelings. Categories and labels are important—it's difficult to imagine discussing the topics of this book without them—but we must remember that they exist to serve us, not the other way around. When we find ourselves boxed in by restrictive categories, there's just one thing to do: break out of them.

A Question of Labels

There's no question that we are living in a time of great upheaval. Societies around the world are bitterly divided along social and political lines, on issues from the environment, immigration, globalism, and the list goes on. But according to many, the most dangerous threat of all is the increasing abandonment of traditional notions of sex, gender, and sexuality and the resulting collapse of the traditional nuclear family.

When I use the term *traditional*, I simply mean the views that have dominated most of the world in the recent past. For generations, we were taught that biological sex is a simple matter: that men are men and women are women; that men should be attracted only to women and vice versa; that romantic dating was aimed at the goal of finding a spouse as quickly as possible; that sexual intercourse should occur only between married couples and with the main goal of making babies; and that marriage should be permanent and focused primarily on the raising of children. These were not presented as aspirational ideals but, rather, the *correct* way of things. Deviation from any of this, we were told, was aberrant, disordered, sinful, and *unnatural*.

The times they are a-changing. More and more of us, especially young people, are taking an open-minded approach to gender and sexuality that goes far beyond identifying as gay or transgender, the pioneering deviant groups of the twentieth century.[1] Terms such as *gender fluid* and *gender nonbinary* are gaining momentum, and young adults are increasingly identifying as pansexual, demisexual, or even asexual, rather than plain old straight, gay, or bi.[2] Conventional categories for gender identity and expression, and sexual attraction and romanticism, are just not cutting it anymore.[3] The ever-growing litany of categories for gender and sexuality are giving way to a free-for-all called *no labels*, an approach that eschews any restrictive definitions whatsoever.[4,5]

Sexual and romantic relationships are changing as well. Many young adults have no intention of getting married,[6] consensual nonmonogamy is on the rise,[7] and birth rates are plummeting.[8] In just two generations, the sexual landscape has completely changed throughout most of the developed world, and so it is no surprise that many people find all of this terribly unsettling.

What *is* surprising to me, however, is how little the biology of sex has factored into the public conversation. After all, sex, gender, and sexuality are, first and foremost, biological traits. Of course there are psychological, social, political, and even legal ramifications for these issues, but I think we have all heard plenty from those perspectives already. It is time for biology to have a seat at the table, and with this book I am forcefully pulling up a chair.

I assert that this moment of sexual turmoil is actually a *rediscovery* of the much more expansive relationship with sex that our ancestors once had and that other animals enjoy today. I use the word *ancestors* to refer to both earlier human societies of our prehistoric past (thousands of years) and the primate ancestors of our deep past (millions of years). To make this argument, I will explore what we observe in other animals and diverse contemporary human societies, as well as what we can infer from our forebears in the ancient world.

The key to taking an unbiased look is to set aside preconceived notions the best we can and look deeply with an open mind, as I have tried to do here.

Implicit in this thesis is the notion that much of how we humans express our sexuality and how we form our sexual relationships stems from cultural constructions, not innate biological wiring. As we will see, even a cursory glimpse of the sex lives of other animals demolishes any notion that sexual activity is narrowly purposed toward procreation. Biologists have discovered an ever-expanding list of reasons that animals have sex with each other. Animals use sex for bonding, social cohesion, and alliance building. They use sex deceptively, competitively, and financially. They even have sex for the same reason that *we* most often do: just for the fun of it.

Moreover, in no other species but humans is sex rigidly restricted by broad constructions such as heterosexuality and sexual monogamy, nor was it so restricted for most of the history of our own species. A reimagining of those constructions doesn't seem so crazy when we consider how peculiar and recent they really are. Animals masturbate. They enjoy oral sex, "gay" sex, and group sex. They engage in "extramarital" sex and struggle to manage the jealousy that often accompanies it. As humans explore this supposedly new sexual territory, perhaps we can look to our animal cousins for guidance on how to manage this minefield of social upheaval. I'm not joking!

Traditionally, the question of why we have sex, and with whom, was answered more by cultural values than the biological underpinnings of this behavior. In my view, it is long past time to bring science into this discussion, and there are at least two ways in which the field of biology can teach us about ourselves. First, a look at the natural history of sex helps to place this interesting behavior in its full evolutionary context. While it is not my claim that the behaviors of non-human animals are the sole or even the best foundation for moral reasoning, studying animals can demystify our own behavior and reduce harmful taboos. And second, by fully understanding the various reasons *why* animals have sex, and what they gain from it for themselves and others, we can more fully appreciate its many social facets and help contextualize sex among our other needs.

Studying the biology of animal behavior is difficult work and requires continual probing and questioning. Understanding human behavior is even more so, given how much more complex our culture is. In some cases, I will draw a direct line from the behavior of close animal relatives to our own, arguing for shared evolutionary history. In these cases, we really can infer a lot about the nature of certain behaviors by understanding how they work in other animals.

In other cases, the line is parallel but not direct, meaning that a behavior may accomplish similar goals, though it may not have been inherited from a common source. Biologists frequently encounter traits that are shared in various organisms, not because the species are closely related but because the traits emerged independently because of their distinct advantages. For example, both birds and bats evolved wings and the ability to fly, but they did so separately. Because of that, their wings have major structural differences, yet they function similarly because they have the same purpose. The same can be true for behaviors. The tendency of present-day humans to form *dyads* (stable groups of two for raising children) is also observed in Diana monkeys. But none of the apes to which we're more closely related, nor any of the monkeys to which Diana monkeys are related, form dyads. Therefore, this tendency to pair up, as it were, developed independently in these two species and we must exercise caution when, for example, we study Diana monkeys as a correlate of human marriage, as some psychologists have. I will point out the shared evolutionary history, or not, as we go.

What This Book Is (and What It Is Not)

The first five chapters are a comprehensive exploration of the biology of sex, gender, sexuality, and sexual relationships in non-human animals. In order to truly appreciate our own relationship with sexuality, we need to place it in its full evolutionary context because our species didn't suddenly fall from the sky. We share kinship with all other animals—the other African apes are our cousins separated by just a few million years of distinct ancestry—and our bodies, our minds, and our behaviors took shape under the same watchful eye of natural selective forces. By observing how other animals engage their sexuality, we gain a much more complete appreciation of, for example, what gender really means, what sex is really for, and why sexual relationships are so important to us. Once we have done that, we are ready to focus on human sexuality, and that's what we will do in the final three chapters.

I approach this topic first and foremost as a scientist. My research is focused on human evolution and the genetic basis of human uniqueness. Specifically, I study how human genes are regulated in unique ways compared to those of other apes. When it comes to anatomy and behavior and all of that, we know a lot about how humans are different from other animals, and how modern humans are different from more archaic species such as Neanderthals, *Homo erectus*, and so forth. But we know embarrassingly little about the genetics underneath those differences, so this is why my laboratory explores human uniqueness from the genetic point of view. We hope that by discovering instances of human-specific gene expression, we can retrace how a species as peculiar as ours evolved.

In 2012, I decided to write a book on human uniqueness. I dedicated my sabbatical year to doing research and reading on the topic, only to find myself writing the exact opposite book that I intended to. In studying how animals behave, and the instincts and emotions that drive those behaviors, I concluded that animals are *Not So Different* from us, and that became the title of the book that resulted from that effort.[9] Similarly, the book you are now reading emphasizes our similarities to other animals rather than our differences.

It may seem confusing that a scientist who studies what makes humans different is constantly writing books and articles about how humans are *not* different, but that's sort of the point. To really understand what it means to be human—the driving question of my life's work—we must explore both our continuity with other animals and our distinctiveness. Like two sides of the same coin, considering either perspective on its own gives you only half the story.

For example, a few years ago, my research group started a hunt for human-unique genes—that is, genes that we have but our closest living relatives (chimpanzees) do not. These species-specific genes are sometimes called *orphan genes* or, in technical terms, *taxonomically restricted genes.** My group took a somewhat naïve approach. We employed powerful new computational

---

* I prefer to call them *young genes*, because I think the mechanisms used in the creation of new genes are far more interesting than which close relatives do and do not share the genes.

techniques for genome analysis that simply aligned human and chimp chromosomes and looked for differences according to precise criteria that we could customize in myriad ways. In truth, I didn't actually expect to find anything new, as other researchers have taken similar approaches over the years. But I thought it was a good way to take this new analysis technology out for a spin and see what it could do.

We chose the smallest human chromosome—#21—with the hopes of learning our way around and then proceeding to more meaty chromosomes. To my great surprise, within a few weeks we had found an island of human-unique DNA, and within this island were some particular types of genes called *microRNA genes*. We spent another two years conducting painstakingly detailed analyses to characterize these genes, infer their origin, and ensure that we were correct in what we thought we were seeing, finally publishing the discovery in 2021.[10] We are now using this approach in a wider effort to discover any and all human-unique RNA genes throughout the human genome. It is an incredibly exciting time to be a genome scientist!

That work, in part, was the inspiration behind my second book, *Human Errors: A Panorama of Our Glitches, from Pointless Bones to Broken Genes*.[11] In that book, I focused more specifically on human evolutionary history and, in particular, the consequences and downsides of our unique history as a culturally evolved—and not just a biologically evolved—species. We are a strange and eccentric species, not just because of our cells and genes, but because of the way that we *create* our own living environment rather than simply react to the world around us.

I mention this because it demonstrates my belief that to truly understand human nature—in this case our *genetic* nature—we must study both the genes we *share* with other animals and those that are uniquely ours. If I had assumed that only shared genetic elements could be functional—a very common assumption in evolutionary genetics—I would have missed those human-specific genes on chromosome #21, like scientists had before. But if I ignored the fact that our chromosomes have a shared history with the chromosomes of chimpanzees and gorillas, there would have been no way to find those genes in the first place.

The same is true for other aspects of human nature, including our social and sexual nature. Parsing out what we share with other animals is as important as acknowledging everything that makes us different. We are a unique species. But we are also one of several species of African ape. Both our shared and our unique history make us who we are. This book focuses mostly on our shared history with other animals, simply because others have already written books about the unique history of human sexuality. The final three chapters summarize some of that work, as an attempt for me to bring us back to the human experience of sex and gender after having spent five chapters discussing other animals.

Most biologists study animals for their own sake and on their own terms, not as a lens for understanding humanity. After all, humans are not the only species with distinct features. *All* species are unique and must be understood as such. Humans are not simply smarter versions of chimpanzees, and we must not reduce other animals to being just simpler versions of ourselves. That being said, when we observe how various animals engage sex, gender, and sexuality, it is impossible to miss certain patterns and commonalities that reveal deep truths about what sex is truly all about. In fact, I would argue that if we want to truly understand human sexuality, we *must* consider the full picture, which necessarily includes the comparative perspectives.

Until we can hop in a time machine and directly observe our ancestors, the best we can do is to observe other living animals as closely as we can and attempt to draw inferences about shared features and how those features play out in the unique environments in which our ancestors evolved. This is an imperfect process, to be sure, but, combined with what we have learned from fossils and artifacts, it has produced invaluable insights into why we are the way that we are.

Be warned: if you are attached to your own prejudices about gender or sexuality, this book may unsettle you. In these pages you will learn that other animals have been experimenting with gender expression for millions of years. You will see how intersex bodies are nothing new or scary. You will appreciate that sexual activity is not and *was never* solely about procreation. You will learn why sexual monogamy is nearly unheard of throughout the animal kingdom. You will grapple with the many human societies in which our contemporary sexual taboos had no place. At the same time, you will be amused at early attempts to explain homosexuality in humans. You will be haunted by how transgender individuals navigated a world that was ill-suited for them.

And you will wonder *just how the hell we got things so wrong.* Just two generations ago, we literally *demanded* that teenagers agree to marry someone, permanently and exclusively, without having even seen them naked, let alone developed an intimate or romantic attachment to them.

To be sure, exploring the sex lives of animals does not provide all the answers. In fact, I would not trade sex lives with very many other animal species. (Except bonobos. If you could choose what species in which to be reincarnated, you'd be crazy not to choose bonobos.) And while there may be enviable aspects of the experience that other animals have, and that our ancestors once had, there are horrors as well. For example, when a male gorilla first takes over a harem, he promptly slaughters all of the children of the previous alpha male. While there is a clear evolutionary benefit to his doing so, this is not an aspect of gorilla life that I would urge us to emulate. Similarly, neither species of chimpanzee recognize any taboo against sexual experiences with young juveniles, even infants, but this does not lead me to reconsider my revulsion toward pedophilia, even a little bit. Just because animals do something doesn't mean that we can or should also.

Nevertheless, when it comes to a more expansive, diverse, and flexible attitude toward gender and sexuality, we can learn a lot by looking beyond our own species. We don't have to emulate anything that we don't want to. To be honest, even *I* am sometimes unsure what to think of all of this. But what I do know is that many of the arguments that have been used to enforce the rigid and limited view of sex that has dominated most of the world for the past few centuries fall apart very quickly when we take a look around. Our biology does not dictate a strict gender binary. Heterosexuality is not the primary sexual orientation. Sexual monogamy is virtually unheard of among mammals. And the nuclear family is not the sole natural state for humans.

You may identify strictly as a woman. Or you may be sexually attracted only to women. You may prefer a traditional approach to dating, relationships, and marriage. You may not want anything to do with a polyamorous relationship. There is nothing in this book that should challenge your personal conviction on those issues. But I do hope you find good reasons to accept that some of your neighbors feel differently and that their views are just as valid as yours. What we see when we look to other animals, or to our own past, is a whole lot of diversity. There is no one right way to be a man, no one proper way to have sex or form relationships, and there isn't a rigid template for how to make a family.

Some critics of the ideas expressed herein will claim that they are driven by a social agenda, an expression of crazy postmodern values, or perhaps even my own personal beliefs or preferences. That is not accurate. When we encounter sexual behaviors in other animals, and even in some past or present human populations, that run against my values, I will have little hesitance in saying so. For example, sexual coercion and forced copulation are not uncommon in other species, even our close relatives, and with both sexes among the perpetrators and victims. Yet I am as staunchly against rape and sexual assault as anyone else. Believe it or not, this book is not about values; it is about biology.

That said, I will not insult you by claiming to have no dog in this hunt. Besides being a scientist interested in human nature, I identify as a gay man and have been involved in gay rights advocacy since the late 1990s. I am also a participant in the institution of marriage. I attended oral arguments before the Supreme Court in *United States v. Windsor*; I have stared down members of the Westboro Baptist Church; and I got married within months of being granted the legal right to do so in my home state of New York. My husband and I are also raising two children, whom we adopted from foster care. To anyone who would accuse me of wishing to destroy the concept of marriage and family: Why would I want to tear down the very institution for which I fought so hard to access?

Whenever hard science is brought to bear on social and personal matters, even public policy concerns, we must ultimately dissect the "is" from the "ought." Traditionally speaking, science has been concerned with matters of "is." Scientists aim to pursue and describe the true nature of things as they really are. It is the realm of other disciplines—such as philosophy, law, and theology—to pursue how things "ought" to be. But the lines are blurry and the dissection of the "is" from the "ought" is never complete. How can we get to where we "ought" to be without fully understanding where we are and how we got here?

For that reason, as much as I will try to offer this book as a dispassionate look at the biology of sexuality, it will unavoidably straddle the world of "is" and "ought," and it would be disingenuous to pretend otherwise.

The simplest and most widely appreciated example of this within the realm of sexuality studies is masturbation. For many centuries in the Christian world, professionals, including scientists, insisted upon a specific "ought": that humans ought not masturbate because it was sinful and harmful to our bodies and minds. Science, when it was finally applied to this question, showed that nearly all humans (and nearly all animals!) masturbate regularly and there is absolutely no harm in it per se, to body or mind. This forced a large-scale switch in what was believed and taught regarding the "ought" of masturbation. While some religious communities quietly maintain the taboo for the sake of doctrinal consistency, it has been dropped virtually everywhere else, and adolescents are now taught in health class that masturbating is perfectly normal and healthy. So, while science is not in the business of making "ought" statements, the truths that science reveals most certainly do have bearing on the development and maintenance of those statements. What I offer here is a little bit of truth, brought to you by careful observations of the natural world. What we do with that truth is up to us, as individuals and as a society.

This book is most definitely *not* a presentation of my own sense of "ought" when it comes to gender and sexuality. To be perfectly honest, I'm still working that out for myself. I simply think that today's young people have hit on something worth exploring, a more expansive view of human sexuality. Whatever "ought" statements we advance, as individuals or as a culture, I think they should be informed by the best possible science. We can consider a more free and expansive engagement of our sexuality and decide to embrace or reject whatever aspects of it we wish. No matter where we end up, we will be better informed by an honest and comprehensive look at the natural history of sexuality.

Shall we begin?

Chapter 1
Evolution's Rainbow:
Males, Females, and More

The second-highest grossing film of 2003 was *Finding Nemo*, an animated family movie about Marlin, a clownfish, searching for his son Nemo. If you haven't seen the movie, consider this your spoiler alert. The movie opens with the tragic death of Nemo's mother, leaving Marlin a widower and Nemo lost, disoriented, and swept away by ocean currents. The rest of the movie chronicles Marlin's search for his lost son. It is a heartwarming tale about love, loss, and growing up.

If the movie were biologically accurate, the story would have proceeded a little differently. Upon the loss of his mate, Marlin would have transitioned to female. By the time Marlin reunited with Nemo, she would have been his mother.

Clownfish (*Amphiprion ocellaris*) live in a precisely stratified social structure. Like many social animals—although not commonly fish—clownfish have a dominance hierarchy that determines feeding and mating privileges. Within each small group, the hierarchy is determined strictly by size, which correlates perfectly with both age and aggression. At the very top position is the largest fish, who is also the oldest and most aggressive. This alpha fish is always a female and the only one in the group. Therefore, Nemo's mother must have been the alpha of their clownfish group.

The alpha female is surrounded by a small harem of males who are ranked by their size and age. The oldest and largest is called the alpha male, and he is dominant over the others but submissive to the alpha female. Only the alpha male is allowed to breed with the alpha female. All of the others must wait their turn—which is to say, until those ahead of them have either died or transitioned to female. Therefore, because Marlin has a son, he must have been the alpha male.

A clownfish group consists of a breeding pair, plus a collection of replacement males waiting their turn. When the alpha female is lost, the group is not only leaderless but also female-less. What are they to do? Being the largest fish left, in order to take over the harem, all the alpha male must do is switch his sex from male to female. So that is what he does!

Clownfish have almost no *sexual dimorphism*, meaning that males and females are essentially identical, and they even have a bipotential gonad called an *ovotestis*. In a process that takes about a month, upon the loss of the alpha female, the testicular tissue of the alpha male withers away and the ovarian tissue quickly matures and snaps into action. Meanwhile, all of the other males move up in rank and begin to grow. The beta male becomes the new alpha male; the gamma becomes the beta; and so on.

By the time Nemo's father, Marlin, found him, she would have completed her transition to female. The two of them would then have found a nearby anemone in which to make a new home. Assuming no larger males joined them, Nemo and Marlin would then have begun to breed. It's probably a good thing that Pixar wasn't striving for biological accuracy.

Let's Talk About Sexes, Baby

It's a rather unfortunate linguistic situation that the word *sex* has two separate meanings and that both meanings often come into play in conversations involving either topic. Of course, *sex* refers to the behaviors and actions associated with sexuality, usually engaging the genitals and involving physical arousal of one or more parties and sometimes leading to procreation. We call that *having sex* and that's the fun part, both scientifically and, well, otherwise. But *sex* also refers to maleness and femaleness. In everyday speech, we often use the term to express whether someone is gendered a boy or a girl, as in "Do you know the sex of the baby?" But the scientific definition of the word, often expanded to *biological sex*, has been progressively narrowed to the point that most biologists use it to refer only to the generation of sperm or egg cells. And how often does that really come up in conversation? Not much. But for our dissection of human sexuality, it is where we must begin because every body part and every behavior in the realm of sex, gender, and sexuality began with the emergence of these two very peculiar types of cells, the sperm and the egg, around *two billion* years ago.

When biologists consider the *sex* of an animal, we are usually referring to which gamete its body makes. *Gamete* is the word for the very specialized reproductive cells that come in two versions—large and small, casually called *eggs* and *sperm*, respectively. Gametes are very unique cells in the body, the only ones with precisely half the DNA that all other cells have. This is key for the fancy genetic dance that occurs during sexual reproduction, and it's the only reason that these cells exist at all. Whenever two gamete cells fuse together as one, both cells will donate their entire complement of DNA to the fused supercell. To avoid doubling the DNA content every generation, our cells cut the amount of DNA in half when we create the gametes.

For this reason, every sperm and egg cell has half the number of chromosomes that the other cells in that organism have. In humans, gametes have 23 chromosomes, while every other cell in our body has 23 *pairs* of chromosomes, one each from our two biological parents, for a total of 46. When we make our own gametes, each one has a random sampling of DNA that originated from our fathers and mothers, but it is crucial that each gamete has precisely one of each chromosome, for a total of 23.

Like us, our parents had two of each chromosome, one from each of *their* parents, but they only pass one of them to us. It's a coin flip which one, and with 23 paired chromosomes, that's 23 coin flips. That comes to more than *8 million combinations* of chromosomes (2^{23}). And we have two parents, so we can double that number.* And on top of this, each sperm or egg will harbor a number of novel mutations. In humans, that number is between 50 and 200. The creation of gametes has diversity-generating steps built in so that unique genetic combinations are created automatically and unavoidably every time

---

* Truth be told, it's far more combinations than that because the two versions of the chromosomes actually shuffle DNA between each other during the formation of sperm and eggs. The number of possible combinations of paternal and maternal DNA in a gamete is essentially infinite.

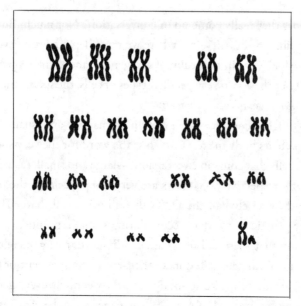

A human karyotype showing 23 pairs of chromosomes

a new individual is created. This is why two parents could have millions and millions of children without any two of them being the same genetically.* The deck is always shuffled.

Because biological sex is determined by which gamete an animal makes and there are two possible gametes—sperm or egg—many people consider biological sex to be a simple either-or binary, even if they accept that *gender* is not so simple. We will discuss the complexities of gender in the next chapter, but it turns out that there are problems with considering sex to be a pure binary as well, even if we use it to identify only which gamete an animal makes. At a minimum, there are *three* sexes because we can't forget about the hermaphrodites. If you thought that biological sex was a simple matter, buckle your seat belt, because this is going to be a bumpy chapter.

---

*   Even monozygotic twins, who begin with nearly identical genomes, develop their own novel mutations and will begin to experience different epigenetic imprinting (chemical modifications to DNA that affect how genes are expressed) very early during embryonic development. As anyone who has ever been close to a pair of them knows, monozygotic twins are *not* identical, even genetically.

Hermaphrodites are animals (or plants) that can make both sperm and egg, often at the same time. While most vertebrates—animals with backbones—feature only males and females, many species of invertebrates—animals without backbones—also include hermaphrodites in their ranks. Some species consist entirely of hermaphrodites; some have hermaphrodites alongside males and females; while still others have what are called *sequential hermaphrodites*, individuals that switch sex at some point in their life cycle.

And then we also have parthenogenic species: animals that can reproduce all by themselves, but not through cloning. *Parthenogenesis* is the term for when an egg develops into a complete organism without being fertilized. A famous example of this occurred in 2006 when a lone female Komodo dragon (*Varanus komodoensis*) in the Chester Zoo in the UK hatched a brood of eggs without the input of a male. The hatchlings developed into perfectly normal Komodos, and scientists have since observed this rare phenomenon in wild Komodos as well.[1] Parthenogenesis has also been observed in sharks, domestic turkeys, and a variety of snakes and lizards.[2] Parthenogenic individuals are female because they produce eggs; the reproduction is asexual, since only one individual is involved; but the offspring isn't exactly a clone, since its genome isn't identical to its parent.

In fact, there are several species of whiptail lizards (genus *Aspidoscelis* and *Cnemidophorus*) that reproduce *only* through parthenogenesis.[3,4] The species are all female and lay eggs that develop into adults without ever being fertilized. Nevertheless, these lizards still have sex!

Because whiptail lizards are *induced ovulators*, they must be mounted in order to release eggs. Thus a pair of females will take turns mounting each other in order to bring them both into ovulation, after which they will lay eggs that develop into daughters. We will say more about these whiptails in chapter 3, but for now, suffice it to say that the notion that sex must be binary because there are only two kinds of gamete is demolished by the existence of hermaphrodites and parthenogenetic animals.

Further still, in recognition that maleness and femaleness are concepts that can apply to organs and structures all throughout the body, not just the gonads, the term *gametic sex* is now coming into use. This helps clarify things for individuals with some body parts that do not match with the sex of their gonads. There are animals—and people—whose bodies are sexual mosaics of typically masculine and feminine forms, and so referring to their gametic sex clarifies which gamete they make.

When we get down to brass tacks, gametes are everything in the grand scheme of evolution. Individual organisms, including each one of us, don't matter at all. We are but blips in an ancient timeline. Our bodies will eventually vanish without a trace, but what may truly live on indefinitely are our genes. We are merely vessels for propagating genes, and gametes are the way that we do that. To really understand gametes, we have to zoom way out on this question and ask something even more fundamental about our reproduction: *Why do plants and animals have sex at all?*

In sexual reproduction, two individuals come together to combine genetic material into a new individual, but this is not the only way that living things can reproduce. In fact, the organisms that are most abundant on this planet *by far*, both in biomass and in the raw number of individuals, do not reproduce using sex. I'm talking about bacteria.

Bacteria and most other microorganisms reproduce through various versions of cloning. One individual, by itself, splits into two (sometimes more) and the new individuals are clones of both each other and the parent cell, which ceases to exist once it splits. This is called *asexual reproduction*, and it is very quick and efficient. The bacterium *Escherichia coli*, the most thoroughly studied organism in the world (and the chief component of human feces), has a doubling time of just twenty minutes. If we assume a suitable broth, warm temperature, and unlimited nutrients, a single *E. coli* can produce one million offspring in just twenty generations—in under seven hours. In fact, if we keep going, again assuming unlimited resources, a single *E. coli* could theoretically produce a ball of offspring that is larger than the planet Earth in just over twenty-four hours. Asexual reproduction is fast and prolific is what I'm saying.

Bacteria do have sex; it's just not connected to reproduction. During a process called *conjugation*, one bacterium can send a few genes to another. Following this exchange, no reproduction has taken place and the two bacteria go their separate ways. The donor is no different for having participated in the exchange, but the recipient now has extra copies of some genes and those copies may not be exactly the same version of the genes the bacterium originally had. They could even be new genes altogether. This bacterium will, in a process that appears purely random, select some of the new DNA to incorporate into its own genome. In this way, bacteria do indeed exchange genes, which contributes to their rapid evolution. For bacteria and many other microbes, sex changes the genetics of *the participants*, rather than creating new individuals with new combinations of genes. When it comes time to reproduce, however, they simply clone themselves.

Sexual reproduction, on the other hand, is much slower and more energy-expensive than asexual reproduction, and it presents a whole host of challenges that asexual reproduction does not. The first challenge is how to blend the genetic material of two separate individuals. This is where the gametes come in. Genomes are huge and contain thousands of genes connected together in strings of DNA, the chromosomes. In order to create a whole new combination of genes, one must carefully sort out the starting material because accidentally missing one gene, or grabbing too many copies of another, can be devastating. Not only that, cells are filled with complex machinery that is interconnected in a complex web of molecular attachments. There are some structures that a cell must have only one of, such as the nucleus and the centrosome. Simply taking two cells and fusing them into one big one is not going to produce a functional cell, even if you figure out how to recombine the genomes safely and correctly.

This is the reason that scientists think that plants and animals ended up with the ingenious system of the sperm and egg. One cell, the egg, is large and contains everything that the new potential organism needs to get started in life. The other cell, the sperm, is tiny, about 3 percent the size of the egg (in humans), and instead of fusing, the sperm cell basically just squirts its chromosomes into the egg cell and that's pretty much it.* For simplicity's sake, just about every biology book (including this one!) claims that sperm and egg "fuse," but in reality, sperm cells are like tiny delivery packages for chromosomes. Fertilization is less the fusion of two cells and more the delivery of chromosomes from the sperm to the egg. Once that happens, the egg becomes a zygote and begins the process of development.

---

* In animal cells, the sperm cell also donates a small structure called the *centriole*.

The blending of two genomes into one creates a new combination of genes, but in order to do this without doubling the DNA content of the organism every generation, we produce gametes with 50 percent of the normal amount of DNA. However, merging two genomes into one would be an impossibly complex sorting process if we had just one of every chromosome. So, to avoid having to keep track of each one, plants and animals have *two* copies of every chromosome (except in the gametes, which have just one). When two of them fuse, the proper amount of DNA is then restored. This system of halving the DNA to make gametes, then restoring the full amount when they fuse, creates unique genetic diversity in every individual that is created.

The use of one copy of the chromosomes in the gametes but two copies of the chromosomes in the full body was a key step in the evolution of sexual reproduction. Bacteria don't have two copies of their chromosomes because they don't have sex. And that's also why they don't need sperm or eggs. Their system is much simpler, faster, and more efficient.

So, since it's slower, more expensive, and more complicated, why do plants, animals, fungi, and other complex life still prefer sexual reproduction?* There are two key advantages of sexual reproduction, and both have to do with diversity.

First, the phenomenon of *diploidy*—having two versions of every chromosome—provides a "backup" copy of every single gene. This means that organisms have a buffer if a gene gets damaged. The other copy can serve as a template for repairing the damage. This buffer makes diploid organisms more robust in the face of mutations. Mutations are changes to DNA that occur because of damage, degradation, or copying errors, and they are usually bad for a gene and harmful to the organism. In fact, diploidy probably first evolved as an escape from the accumulation of harmful mutations in primitive microorganisms that were rapidly evolving on the early Earth.

Having a backup copy of every gene is not just great for correcting harmful mutations. It also allows the freedom to experiment with *beneficial* mutations. While most mutations are harmful or neutral, every once in a while, a mutation can be helpful or creative, and these are the raw materials of all evolution-

---

* Although many *can* employ asexual reproduction under certain conditions.

ary change and innovation. Sexually reproducing species, therefore, have more flexibility to experiment with different versions of genes since all of their genes are paired with a backup copy that can retain the original function. It's like if you have only one bicycle. You're not going to risk ruining it by attempting fancy modifications. But if you have an extra one, you can go to town trying all kinds of "mods" because if it doesn't work out, you still have the unmodified bike to fall back on. In organisms like bacteria, there is no compensation for a harmful change, so mutations are much more dangerous and they have far less opportunity to creatively experiment.

Being tolerant of mutations allows sexual species to accumulate genetic diversity, and this is powerful even when it offers no immediate advantage. Every organism can have two versions of every gene, and in a sexually reproducing population, there can be hundreds of different versions circulating in the gene pool, creating a huge reservoir of genetic diversity. Diversity means that each organism has unique traits and features, different strengths and weaknesses, and as any stockbroker will tell you—diversity is the best insurance plan against an uncertain future.

Second, by linking sex and reproduction together, new *combinations* of genes are ensured. Because asexual reproduction is so incredibly fast and efficient, organisms would be tempted, so to speak, to fall back on this incredibly prolific mode of reproduction. In a head-to-head competition, cloning quickly overwhelms sexual reproduction, so an organism that is successful and well-suited to its environment can just make more of itself and continue being successful. But success would come at the cost of diversity, and the big problem is that *success is temporary*. Environments are in constant flux, and stability is short lived. Even the most well-suited organism becomes vulnerable whenever conditions change. A population of clones all live and die together, but a population with a great deal of diversity has a fighting chance of surviving an environmental change, even a catastrophe, because somewhere within that diversity are some individuals who can ride out the storm and lead the species to adapt to a changing world.

The tendency of nature to generate diversity can be seen all around us. From ladybugs with their spots and background colors reversed, to the rare all-black variants seen in North American squirrels, diversity appears constantly. When a spontaneous change first appears, we call it a *mutant*. If it persists for a little while as a minority trait, we call it a *variant*. And if it becomes a major and enduring part of the variation of a species, we call it a *morph*. The appearance of these variations is the work of random molecular tinkering, not sex per se, but sex is key to spreading the variation through the population and ensuring that it is unlinked from other traits so it can be selected, for or against, on its own merits. For squirrels, black coloring may provide better camouflage in heavily wooded environments, but the standard brownish gray works better in more open landscapes. By having the genes for both morphs circulating in the population, squirrels are more adaptable.

Sexual reproduction forces a species to continuously generate diversity, spread it through the population, and resist the allure of cloning. The biological lesson here is that diversity wins, especially in the long term, and it always has. Because sex is the factory of diversity, it's only fitting that, at long last, we are now exploring the diversity of sex itself.

We often take the human style of sexual reproduction for granted, as though it couldn't possibly be any other way, but it turns out that there are lots of ways that organisms can enjoy the benefits of sex—that is, genetic recombination—and our egg-sperm way is just one of them. That said, the system of one large gamete and one small gamete offers the best balance of the pros and cons of sexual reproduction, so there are good reasons why all plants and animals have settled on it. But first, let's take a quick look at some of the options that I think we're all glad were mostly discarded by our ancestors.

Sugar and Spice and All Things Nice

There are creatures called slime molds and water molds that, to be perfectly honest, are just as gross as their name implies. For example, there is a species of slime mold, *Fuligo septica*, whose common name, I kid you not, is the "dog vomit slime mold." These disgusting organisms are not actually true molds and, in fact, are in their own kingdoms, which is definitely where they belong, given how strange they are. But they exhibit fascinating reproductive diversity, a callback to the primordial Earth, as life was experimenting with various modes of sexual reproduction.

For instance, a slime mold called *Dictyosteliida* exists for most of its life as a single-celled life-form that looks like the amoeba you may have seen under a microscope in high school biology class.[5] Most often, these single-celled creatures reproduce by asexual cloning. However, when food or water becomes scarce, they do something quite remarkable. The individual cells all send out a "homing signal" to attract each other and then they get together to form a large mass of cells resembling a small hairy slug. This mass of cells can even move around in a coordinated fashion, as though it were a multicellular organism (reminiscent of the Mind Flayer from *Stranger Things 3*). From within this structure, some cells fuse to create spores—a form of sexual reproduction. Other cells sacrifice themselves and join together to form thin stalks, or fruiting bodies, that will release the spores, like a mushroom or the fuzzy blue part of a bread mold. In other words, thousands of distinct genetic individuals come together to create a giant sex organ that spews out spores. I guess that's one way to do it.

In an almost exactly opposite example, any discussion of the diversity of animal reproduction would be incomplete without mention of the bdelloid rotifers (class *Bdelloidea*).[6] The bdelloids (the *b* is silent) are tiny, microscopic animals, and they comprise an ancient lineage descendant from the very first spongelike animals. They live in fresh water, are completely transparent, and have the distinction of being the only major animal group that has done away with males altogether. These all-female creatures produce eggs that then grow into new organisms all by themselves. Because there is no sex, there is no genetic recombination and thus very little diversity.

The fact that the bdelloids have persisted to the present day with no way to ensure genetic diversity is astonishing, but two recent discoveries have shown us how they have managed this.[7] First, these animals thrive in a harsh lifestyle that cycles between an aquatic and a completely dry environment that few of their parasites and competitors can endure. Among the most pressing reasons why organisms must pursue genetic diversity—and perhaps the initial reason why sex evolved in the first place—is to outrun and outwit parasites and pathogens. Therefore, since bdelloids can shake off their parasites by cycling between two extremely different environmental habitats, genetic diversity just isn't as vital for them as it is for the rest of us.

Second, this group has benefited from a massive influx of new genes via *horizontal gene transfer*, through which genes from unrelated organisms are transferred during viral infections.[8] When viruses move around through a population, they can inadvertently bring genes with them, and when viruses jump from one species to another, they can accomplish, purely by accident, what can never occur otherwise: the exchange of genes between two species without any interbreeding. This is extremely rare, but when it happens, it can lead to remarkable evolutionary adaptations. So bdelloids are actually the exception that proves the rule that sexual reproduction really is the way to go. Short of the extraordinary ways that bdelloids have managed to compensate, asexual reproduction is a recipe for extinction among animals.

If we consider the bdelloid rotifers an all-female species, creatures such as fungi don't really have "sexes" at all; they have *mating types*. Many fungi, including baking and brewing yeast (*Saccharomyces cerevisiae*), have two mating types, *a* and *α* (the Greek letter *alpha*). These are not sexes because the two cells that fuse are not different sizes and a different chain of events follows the fusion of two of these gametes. In the case of sperm and eggs, one of each fuse together to create a single unique individual. However, when an *a* and an *α* gamete fuse together, they create a large diploid cell, which then immediately divides to generate a variety of genetically distinct haploid spores, which can then grow into several unique haploid individuals. This system accomplishes the same thing as the sperm and egg in humans—that is, the shuffling of different genes—but it has one big disadvantage: the regular body cells of adult fungi do not have the backup copy of every gene that they can play around with through mutations. This is probably one reason why fungi never came to dominate Earth's landscape the way that plants and animals have. Instead, they are mostly relegated to their undignified but essential ecological role of decomposers.

While some fungi have just two mating types, such as *a* and *α*, or *Mat1–1* and *Mat1–2*, others have four. These are called bipolar or tetrapolar mating systems, respectively. It doesn't really matter what the types are and how they are different—it basically just comes down to molecules poking out from the surface of the gamete cells. The only hard rule is that a cell cannot mate with another cell of the exact same mating type. a and a are incompatible, as are *α* and *α*. For this reason, tetrapolar mating types have a big advantage over bipolar types when looking for a mate. They are compatible with 75 percent of the population of their species, while bipolar mating types are compatible with only 50 percent of their conspecifics. As you might already have guessed, there are some fungi that have taken the "mating type" phenomenon to an obnoxious extreme and have dozens or even hundreds of mating types, increasing the percentage of the population with whom they are compatible to nearly 100 percent.

Currently, the world record belongs to *Schizophyllum commune*, a white mushroom that has over 23,000 different mating types.[9] When this fact was discovered, many press outlets reported that this mushroom has 23,000 "sexes"—or, worse, 23,000 "genders." Um, no. Mating types are not the same thing as sexes and they certainly aren't genders.

The thing to really understand here is that, although alternatives exist, our gamete system (sperm and egg) seems to be the best balance of the pros and cons, and that's why plants and animals have opted for it. Yes, it's slow and it also means that each of us is only reproductively compatible with 50 percent of the population, but it enforces sexual recombination, avoids the temptation of cloning, provides backup copies of genes to allow experimentation, and involves no more than the minimum number of gamete types—two—to allow all of this. The evidence that this is the best compromise is that so many successful lineages have settled on this system. All plants and animals are diploid, meaning they have two copies of all chromosomes, and they have two different kinds of gametes, a big one and a small one.

Back to animal sexes: in most vertebrates, including mammals, there are males, which make the small gamete, and females, which make the large gamete. However, many animal species also include hermaphrodites, especially the invertebrates. While we may think of vertebrates as the "higher" animals, this is a species-chauvinistic position, especially when we consider that invertebrates are older, are more diverse, and far outnumber us on the planet, in terms of the number of individuals and the number and diversity of species. This is an invertebrate world; we just live in it. Hermaphroditism is common among invertebrate animals and nearly universal among the oldest lineages, and it is overwhelmingly dominant in plants as well (although different terminology is used for plants). This indicates that hermaphroditism is older than having separate sexes.

Hermaphrodites have one enormous advantage: they never have to worry about finding a mate, since they can simply mate with themselves, a process called *self-fertilization* (or *selfing*). However, because it does not promote genetic diversity, selfing is usually used only as a last resort, a poor substitute for sex with someone else. (That probably sounds a little familiar, but if you're hoping for a section on masturbation, the bad news is that you're in the wrong chapter. The good news is that you're in the right book. Keep reading.)

Selfing by hermaphrodites is not cloning. Strange as it seems, a hermaphroditic animal that self-fertilizes is engaging in a form of sexual reproduction. What determines whether reproduction is sexual or asexual is not how many individuals are involved, but whether the reproduction introduces the mingling of two different genomes. Asexual reproduction involves only one genome, perfectly duplicated (less any copying errors) and produces clones. Sexual reproduction involves the fusion of two gametes, and it doesn't matter if the two gametes come from the same individual or not.

When any plant or animal creates gametes, each gamete contains a unique combination of the organism's two sets of genes. In a purely random sorting process, some genes will have come from the individual's biological father and some from their mother. Because organisms have tens of thousands of genes, there is a nearly infinite number of combinations. Male animals typically generate *billions* of sperm in their lifetimes, and no two are exactly the same in their combination of maternal and paternal genes. Although females may "only" generate tens or hundreds of thousands of eggs, they are all totally unique in their gene combinations. Therefore, when an egg and a sperm unite, even if they come from the same individual, the resulting offspring is not identical to the parent. Instead, they are a unique combination of their grandparental genes. And they would have only two grandparents, not four.

All of that said, even though selfing is technically a form of sexual reproduction and does involve some genetic recombination, it is not good for genetic diversity. Instead of drawing on the gene pool of a population of organisms, it draws only on the gene pool of a single person. It therefore creates *subsets* of the genetic diversity that the parent has. When gametes from two *different* individuals fuse in a sexually reproducing population, genetic diversity multiplies. But when two gametes from the *same* individual fuse, the diversity *halves*. This is an extreme form of inbreeding, and hermaphroditic plants and animals go to great lengths to prevent self-fertilization.

Flowering plants get really creative with their self-avoidance. A few plants, such as gingko trees, have separate male and female individuals. Others put the male parts in different flowers than the female parts. Still others carefully time the sequential maturation of their male parts and female parts such that they are not receptive to fertilization at the same time. Some flowers even cleverly arrange their male and female parts in a physical configuration such that a pollinator, most often a bee or other insect, is unlikely to transfer pollen from the male to the female parts within the same flower. For example, the parts may be arranged so that a bee does not brush past the male parts on the way *to* the female parts (where the nectar is stored), but only on the way *out* of the flower after it has visited the female parts and drank the nectar. This ensures that the pollen that gets transferred to the female parts is from the previous flower the insect visited, not the current one. It's not perfect, but it usually works.

Hermaphroditic animals just have to restrain themselves. Unlike plants, animals have this crazy thing called "behavior," which is the subject of most of this book. Just like their anatomy and biochemistry, the behaviors of animals are the product of evolution. If behaviors lead to long-term success, they will be selected and tweaked. If behaviors lead to dead ends, they will eventually disappear. That's how natural selection works. Because selfing is such a bad strategy in the long run, hermaphroditic animals have evolved to avoid it at almost all costs. But if an animal is alone and isolated for a long period of time, it is a handy option of last resort for an animal facing its demise.

The Middlesex

There are two main kinds of hermaphroditism in animals: simultaneous and sequential. Simultaneous hermaphrodites are those who are male and female at the same time for their whole lives. Sequential hermaphrodites are those that switch sexes at some point in their life cycle. Sequential hermaphroditism is pretty clever because it brings the flexibility and adaptability of being able to make either gamete, but it avoids the long-term dangers of inbreeding because only one kind of gamete is made at any one point in time, rendering selfing impossible. But first, a word about simultaneous hermaphrodites.

Two earthworm hermaphrodites mating

For most slugs, snails, and worms, all individuals are simultaneous hermaphrodites. Each has separate ovaries and testes and both tissues are productive and fertile for all or most of the animal's life. However, theirs is anything but a solitary existence. These animals are constantly on the lookout for mates. Food and sex are the two main appetites for basically all animals, and slugs and worms are no different. Hermaphroditic species engage in courtship, mate competition, and all of that. In order to have sex, an individual must find a potential mate and convince them to have sex. It's really not any different except that there aren't "sex roles" and everyone is more or less on the same playing field.

Among these invertebrates, hermaphroditic sex even *looks* familiar to us. They arrange their bodies such that the female parts of one individual are lined up with the male parts of the other individual and vice versa. In other words, they "sixty-nine." In typically stoic fashion, scientists call this *bilateral sperm transfer* because both parties mutually inseminate one another in order to fertilize their eggs. Invertebrates don't have pregnancy and most don't engage in any form of parental care, so, following sex, the fertilized eggs are simply deposited in the water or soil and the parents go on their way, hoping the best for their progeny.

It's unclear what kind of perceptions or inner experience invertebrates have, but if sex is something they "enjoy," the way most of us do, then these hermaphroditic invertebrates have enviable sex lives indeed. Not only do they have scores or even hundreds of sexual partners, the sex itself can last for hours, with considerable "foreplay" leading up to the transfer of sperm. Although it's not clear what is going on during these extended copulatory sessions, they almost certainly serve some kind of purpose, given that a squirt-and-go approach would work just fine.

Biologists believe that earthworm foreplay is a form of quality assessment in which each individual is somehow exploring the health and fitness of the other. We think this because sex is not always completed and, instead, individuals can disengage before releasing sperm. Scientists have found that individuals can change their mind and can even attempt to "cheat" by *giving* sperm but not taking it. One might have thought that one giant annoyance that hermaphrodites are exempt from is the "battle of the sexes" that plagues many of us vertebrates, but, alas, they are not so lucky.

In biology terms, the "battle of the sexes" refers to the divergent reproductive interests of males and females and the inevitable conflicts that result. We'll dissect that later on, but you might be wondering how hermaphrodites can possibly be in reproductive conflict when everyone is the same sex. The reason for this is that sperm is cheap and eggs are expensive, owing to the big size difference of the cells. For the cost of one egg, an animal can make hundreds or thousands of sperm. Therefore, hermaphroditic animals are often a bit more choosy about whom they *take* sperm from, but they'll *give* sperm to just about anyone. So the worms we see in those amorous sixty-nines might actually be in competition with one another to see "who's the daddy," so to speak. Of course, the ideal situation would be that both individuals are duly impressed and agree to take the sperm of the other. Short of that, some pretty hilarious situations can ensue.

The favorite hermaphrodites of many biologists are the flatworms in the genus *Pseudobiceros*.[10] Armed with two penises, these marine worms actually compete for the right to inseminate in a spectacularly named dance called *penis fencing*. In bouts that can last up to an hour, these hermaphrodites go at each other repeatedly, each attempting to insert one of their penises inside the vulva of the other. Without any other appendages, the main defensive weapon the worm has is, once again, its two penises. The resulting conflict is like a slow-motion duel, with both combatants brandishing two swords.

Even given the higher cost of eggs, this competition seems a little silly. No matter who wins, both worms will be the parents of the resulting offspring, and eggs are not *that* much more expensive than sperm. So why do they spend an hour literally jousting over the privilege of fatherhood? As we will see over and over again in this book, there is much more going on than is apparent at first glance.

The fight between these flatworms is almost certainly not about the price difference in eggs and sperm. Instead, it is likely a courtship battle, a struggle to see who is a worthy opponent and therefore an ideal mate. It's a quality assessment. Being vanquished in this battle is not actually a "loss"; it means that one has encountered a strong and formidable opponent, someone that any worm would be fortunate to mate with. After all, losing simply means that one's genes will be comingled with those of the victor in a new generation of, hopefully, stronger and more robust worms.

If simultaneous hermaphroditism seems straightforward, the sequential version is anything but. Sequential hermaphroditism is sometimes referred to as sex-switching, and there are species that switch female-to-male, others that go male-to-female, and some that are born as unspecified, or androgynous, and can develop into males or females upon sexual maturity. These are all forms of sequential hermaphrodite, but they have two important gains over simultaneous hermaphroditism. First, there is no temptation for selfing, as mentioned before. And second and most important, these hermaphrodites can respond to various environmental or social conditions and switch sexes if and when it will bring maximum advantage to the individual. That flexibility can really pay off. Some examples will illustrate.

We can start by returning to the matter of Nemo and his father, Marlin. As shown in the opening and closing scenes of the movie, clownfish make their homes inside sea anemones, but what is not mentioned is that anemones are incredibly venomous. Armed with barbed stingers, they can inject such poisonous venom that a single sting from the most dangerous species can cause anaphylaxis and organ failure in humans. Clownfish, however, are uniquely immune to sea anemone venom and impervious to their stingers. The two species live in a beautiful example of symbiosis, with anemones pro-

A clownfish family living in a sea anemone

viding protection from almost any possible predator and clownfish bringing food to the anemones in the form of their leftovers and their waste. Despite the relative safety that anemones provide, clownfish are fiercely territorial, as they defend their home from competitors—that is, other clownfish.

As mentioned above, clownfish families consist of a breeding pair—an alpha female and an alpha male—and a small number of smaller and younger males, often the offspring of the breeding pair, ranked by their size and age.[11] All family members defend against home invasion by other clownfish. Meanwhile, the younger fish stand ready to move up the ladder, replacing individuals that die (or leave). The alpha female was previously the alpha male; the beta male is next in line to become the alpha male, and so on down the line.

Scientists have observed how clownfish genes initiate the sexual transition. The first place we can detect a change is in their brain, as the alpha male clownfish becomes aware that (a) the female has died; and (b) it is now his turn to become the female. Signals are sent from the brain to the gonads, and the transition begins there. For the rest of the males, they simply grow a little bit and move up the dominance rank by one position. As they continue to grow and age, they are patiently awaiting their turn to be the breeding male and then eventually the breeding female. Hopefully they avoid all the dangers of the ocean so that they can actually take their turn when it comes. But most don't make it.

This social structure is pretty peculiar among animals and is worth exploring in detail. One might think that restricting reproduction to only the top two fish would needlessly restrain potential population growth, a poor strategy for success. But we must remember that fish species tend to produce, release, and fertilize *thousands* of eggs per female. Production is just not a limiting factor. What *is* limiting, however, are the sea anemones where the clownfish make their homes. Simply making tons more fish that there isn't room for is not just wasteful, but it could harm the host. Symbioses are delicate and sea anemones are especially fragile. Instead of mass production and competition, clownfish maintain strict discipline in their growth and dominance hierarchy. Instead of direct competition with one another, their method of quality control is simply that the breeding pair has grown, developed, and survived to older age and therefore must be reasonably robust and healthy individuals.

Perhaps most remarkable is that the nonbreeding members of the family appear to control their own growth in order to maintain the precise pecking order. They "know their place," so to speak, and will pause their body growth rather than lose their place in line or overtake an older fish. Especially given the usual evolutionary narrative that animals are always out for their own benefit and look for every chance to outmaneuver their colleagues, this seems truly astounding. Nevertheless, maintaining an orderly progression seems to be part of the natural clownfish discipline.

To demonstrate this, a scientist performed experiments in which he removed a single member of a clownfish family.[12] When he did so, the younger fish who were lower in the dominance order sped up their growth to assume their new rank, while the growth of those that were already larger than the newly absent fish was unaffected. The forces that keep this level of social discipline despite the enormous evolutionary temptation to "cheat" remain a mystery, but the success of clownfish in their unique niche speaks for itself. The system works.

Only a handful of marine animals have evolved to be impervious to sea anemone venom; besides clownfish, only one other species lives in such a tight symbiosis, the anemone shrimp (*Periclimenes brevicarpalis*).[13] Amazingly, anemone shrimp are also sequential hermaphrodites! Besides that commonality, the social structure is entirely different than that of clownfish, and when individuals do transition, it is usually from female to male. But still, this seems to be such a remarkable coincidence that it probably isn't a coincidence at all. The severely limited space in their housing quarters creates a very specific kind of pressure on how they live and reproduce. The flexibility that sequential hermaphroditism allows must be especially useful in these conditions. Life finds a way.

Deep in the ocean hides another type of sex-changing swimmer. There are thirteen known species of bristlemouth fish within the genus *Cyclothone*, and this is believed to be the vertebrate genus with the greatest number of living individuals on the entire planet. With literally quadrillions of them in the ocean, it's sort of amazing that most of us have never heard of them, but this is because they are too small and swim too deep to be of any commercial value. Nevertheless, after centuries of ignoring these fish, in the 1960s some scientists finally got around to looking up their skirts. Because they travel in enormous schools, scientists can easily net huge numbers of them for study, and when they have, they've discovered that the fish are almost all females.

It turns out that most *Cyclothone* species are sequential hermaphrodites and individuals begin their lives as males.[14] When they reach a critical size, however, they switch to female. The reason that scientists were catching only females is that they were catching only the larger adults—and pretty much all adults are female. The smaller males slipped through the nets. This was figured out in laboratory aquariums because we still don't know much about the spawning and early development of these fish in the wild. It probably occurs at depths so great that they are currently out of our reach. Intriguingly, this sex-switching system is responsive enough to allow some fish to remain as males if there is a dearth of males around them. Although it hasn't yet been observed in the wild, this kind of *quorum sensing*, through which a member of a species can detect and respond to the demographics of its group, is very sophisticated and thus unlikely to be a fluke observed only in the laboratory.

Sequential hermaphroditism is not limited to fish. In 1989, scientists from Cornell University published a report that a species of tree frog from West Africa (*Hyperolius viridiflavus*) was able to switch sex from female to male in their laboratory, an observation famously quoted in the 1993 blockbuster *Jurassic Park*. In that study, seven of twelve females changed into males after having laid at least one clutch of eggs, usually more, and all of which went on to successfully fertilize more offspring as males.[15] These sex-switching frogs had offspring in their terrarium, some of whom they were the mother, and others of whom they were the father! Although sex-switching in this species has yet to be observed in the wild, the frogs were not treated with any hormones or chemicals, and similar laboratory observations have been made in at least two other frog species from different parts of the world. Therefore, it seems unlikely to be an artifact of the lab setting.

Both simultaneous hermaphrodites and sex-switchers demonstrate that many animals approach the matter of biological sex more creatively than a simple binary would predict. However, this begs a rather obvious question: If sex is determined by our chromosomes, how can an animal possibly change its sex? The surprising answer is that chromosomes are not always the determinant of sex.

Snips and Snails and Puppy Dog Tails

In 2015, news reports announced that in Australia, climate change was turning female bearded dragons into males. While the reaction from most of the general public was something like "Say what, now?," zoologists were less surprised. This is because it has long been known that sex determination can be affected by the environment, especially so in reptiles. To explore this, we need to discuss how an animal becomes a male or a female.

In mammals, including humans, the story of sex determination begins with the sex chromosomes, the X and the Y. The typical story is that females have two X chromosomes and males have one X and one Y. While these two chromosomes are thought of as a pair, they could not be more different. The X chromosome is huge, the eighth largest of our genome; the Y chromosome is the third smallest, and it has very few genes on it, none of them essential to life—a fact made clear by the fact that, you know, females don't need them at all.

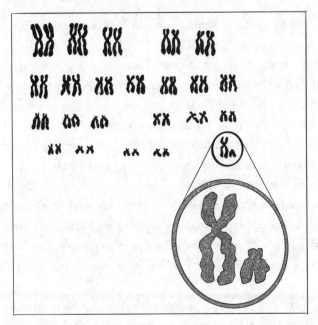

The human sex chromosomes

When we make our gametes, we separate our two sex chromosomes from each other such that every gamete gets just one. Therefore, all of a woman's eggs will have just one X chromosome, either the one she inherited from her dad or the one she got from her mom.* But for a man's sperm cells, half will have an X and half will have a Y. In mammals, the sex of a new individual is thus determined by which chromosome the sperm cell delivers to the egg, an X or a Y. (History holds many stories of powerful men frustrated by wives repeatedly "giving them daughters" when they were hoping for a son, but if

---

* By this point, however, a woman's two X chromosomes are not clean copies of what she separately received from her mother and her father. This is because, as our maternal and paternal chromosomes are preparing to separate, they swap material with each other. After that point, they are shuffled mixes of maternal and paternal genes. In males, however, the X and Y chromosomes are not very similar so they only nominally participate in this process. Fathers pass to their daughters pretty much the same X they got from their mothers. But since females *do* have two X chromosomes, one from their mother and one from their father, they do participate in the shuffle, and the chromosomes that get passed to their children are shuffled versions.

there was any blame to be borne, it would more correctly be placed on the man's own testicles. But I digress.)

The chromosomal system of sex determination is deceptively simple. It's actually an elaborate cascade of gene expression that begins with a single gene found only on the Y chromosome—the *SRY gene*, which stands for <u>s</u>ex-determining <u>r</u>egion of <u>Y</u>. Without the expression of this gene, all mammal embryos will develop into females by default. Popular feminist lore holds that all humans start off as female. That's one way to look at it, but most embryologists would say that we all start off *sex-unspecified* and will develop female unless a signal from the Y chromosome intercedes and sends the embryo in the male direction instead. There is nothing magic about the Y chromosome, it just harbors the SRY gene that can turn an embryo male. Since the presence or absence of the Y chromosome is the key factor, we call this a Y-centered system.

The XX and XY pattern of sex determination is what most mammals use, but there are variations on this theme in other groups of animals. A subtle variation is found in some insects in which the females are XX and males are XY, just like us, but it is not the presence of Y that makes the difference. Instead, it is the number of X chromosomes that determines sex. Two Xs make a female and one X makes a male. This is a subtle but important distinction because in these insects, just like in humans, individuals with just one sex chromosome occasionally appear. When they do, they always have a lone X.* In both insects and humans, lone X individuals are mostly normal and healthy, though sterile. However, in humans they are female while in insects they are male. These insects, therefore, have an X-centered system of sex determination.

Birds, on the other hand, use a similar system but with chromosomes called Z and W. Females harbor a Z and a W chromosome, while males are ZZ. Because all birds and many reptiles use this system, it is believed that dinosaurs did also, since birds evolved from a group of dinosaurs called the *theropods*.

Social insects—ants and bees—also use a chromosome-based system, but

---

* A lone Y is not possible. We all must have at least one X because it has thousands of essential genes on it.

it is totally different. Unfertilized eggs, which contain only one copy of every chromosome, develop into males, while fertilized (diploid) eggs develop into females or sterile males. So, in a sense, their entire genome acts like a giant X chromosome, with one copy indicating male and two copies indicating female.

The award for the most obnoxious system of sex determination has to go to the platypus, a creature covered in fur like an otter but brandishing a bill like a duck. These quirky animals are technically mammals, even though they lay eggs, and are joined by just four other species in a group of very unusual animals called the *monotremes*. Monotremes have *ten* sex chromosomes. Females are XXXXXXXXXX and males are XYXYXYXYXY. This is yet another way that the platypus is so extra.

But chromosomes are not the only way that the sex of animals can be determined. In alligators and some turtles, chromosomes don't matter at all. During embryonic development, there is a critical time period when the temperature at which the eggs are incubated determines the sex of the developing embryo. These species don't have an SRY gene. Instead, they have a combination of temperature-dependent genes whose expression can vary. Some temperature-dependent sex-determined species respond to high temperatures versus low temperatures in differentiating the sexes, while others will vary based on moderate versus extreme temperatures. Because we're so accustomed to the chromosome-based system, the temperature-based one seems odd and prone to potential problems and unintended consequences. But we must remember that, first, it seems to work just fine; and second, the chromosomal system has its own quirks, which we'll explore in chapter 7.

This brings us back to the bearded dragons. These gentle and charming lizards can grow to about the size of a small banana and are kept as pets throughout the world. Like most lizards, bearded dragons employ the ZW chromosomal system of sex determination in which males are ZZ and females are ZW. However, they are also temperature-sensitive. When eggs are incubated at temperatures above 32°C (about 90°F) during a critical developmental period, the growing embryo will become female, regardless of which chromosomes it has.[16] Scientists have called this a "temperature override" and have discovered the precise molecular details of how elevated temperatures affect the expression of a key sex-determining gene called *Jumonji*.* The activation of the *Jumonji* gene by high temperatures initiates a cascade of events that overpowers the sex chromosomes, and it is not currently known how many other lizards share this dual regulatory system.

The purpose of the temperature override isn't understood for certain, but most herpetologists (scientists that study reptiles and amphibians) believe that higher temperatures are most often indicative of an environmental crisis or catastrophe. In such times, having more females is desirable because the number of females is the limiting factor on population growth. Only females can lay eggs, and a very small number of males can cover a large number of females when it comes to fertilizing those eggs. In times of ecological devastation, when populations are crashing, the best way for a quick rebound is to boost the number of females.

Along these same lines, in a variety of plant and animal species, sex determination is influenced by population density. In a species of parasitic roundworms that live in animal intestines—including ours—the ratio of females to males is strictly tied to the overall population size within the parasite's host.[17] At low population densities, hatchlings are predominantly female, in order to rapidly expand the population. At higher population densities, most new hatchlings develop into males. This means that the inevitable competition for limited resources is borne mostly by the males. Because the vast majority of an animal's genes are used in the same way in males and females, the male-dominated competition that occurs at high population densities impacts the entire population and ensures that genes are refined through natural selection.

This particular sex-dependent division of labor is pretty clever and worth saying a little more about, since many different kinds of animals use it. Even in species without pregnancy or social-parental care, females usually invest more in offspring than males do because of the much higher cost of producing eggs than sperm.** Therefore, in some species, females are the primary engines of

---

* Geneticists, particularly those that study non-human organisms, have a penchant for giving eccentric names to genes that they discover. This is just one of our many lovable qualities.

** The higher cost of eggs led to the formulation of *Bateman's Principle*, which took this idea much further and declared that females always invest more in offspring than males do, resulting

reproduction, while males bear more of the burden of competition and quality control. When numbers are sparse, the population is in danger of dying out. This calls for prolific reproduction, which requires as many females as possible. However, when the population is flourishing at high density, this is an opportunity for improvement through competition and natural selection, and that's when the species that are capable of doing so switch to producing more males than females. Although hardly universal, the burden of reproduction falling more on females and of competition being borne more by males is a common theme in animals.

Freshwater eels have a population density–based system of sex determination that is even more elaborate.[18] Juvenile eels remain sex-unspecified right up until they approach sexual maturity, at which point their gonads develop into either testes or ovaries. Through a series of elegant experiments, scientists have discovered that the rate of growth appears to be the determining factor in sex determination. Eels that grow very quickly through childhood develop into males while slow growers develop into females. These differences in growth rates are most often affected by population density with, once again, dense populations favoring rapid growth and, thus, more males. However, other factors have influence as well, including temperature, salt concentration, and the richness of food resources. This provides multiple environmental "inputs" to which the species can respond. By being flexible with their sex determination, eels can fine-tune the ratio of males to females that best suits the conditions.

Unfortunately, as with many invertebrates and amphibians, the presence of estrogen-mimicking industrial pollutants called *xenoestrogens* is disrupting the normal controls on sex determination in many eel species, making them less responsive to environmental cues.[19]

---

in a battle of the sexes in which males want to have as many sexual partners as possible and females prefer just one good one. We now know that this was vastly overblown and overgeneralized and was supported by chauvinism more than scientific data. But it is true that reproductive costs are higher for females than for males in the majority of animals, and this does have consequences for the way that natural selection presses against the two sexes, as explained in this section.

•

The summary here is that sex determination is more complicated and diversified than it seems and the animal kingdom is teeming with sexual diversity. As we will see with gender in the next chapter, in most animal species, sex is a dynamic and malleable concept that can often respond to environmental conditions as needed. On the surface, it seems that humans have opted out of all of this variation and diversity in favor of a simple chromosome-based system. But while that may be true for most of us, the existence of a significant proportion of individuals we know as *intersex* argues that our species has managed to keep sexual diversity alive and well, in the event that it may be needed in the future. We will hold off on discussing human sexual diversity until chapter 7.

Until then, consider the diverse approaches to biological sex described in this chapter. From the penis-fencing flatworms to the sex-switching clownfish, these interesting variants all began with a subtle tweak caused by a mutation. What could have been described as a defect or aberration was actually the beginning of something new and creative. In my view, this perspective should inform our approach toward people with intersex differences. They are not defective; they are merely different. Nature loves diversity. We should too.

Chapter 2
Bending Gender

While *biological sex* refers to the body and cells of an individual, *gender* is another story altogether. In humans, the word *gender* has little to do with biology, at least explicitly, and all to do with culture. *Gender* refers to things such as outward appearance and mannerisms, social roles, and internal identity along a spectrum of feminine to masculine.

The term *gender* has only been used this way since the 1950s. Prior to that, the word *sex* was all there was to refer to people, while *gender* was merely a grammatical term and referred to masculine and feminine qualifying words like *he* and *she*, *his* and *her*. Unlike most languages, English doesn't make much use of gendered grammatical declensions anyway. Nouns are usually neuter and articles and adjectives always are. Only explicitly male or female things such as humans and animals are linguistically gendered, and even then, only in the form of pronouns. A few professions are gendered, such as waiter/waitress, actor/actress, comedian/comedienne, but those are now falling out of favor. The point here is that this use of the term *gender* for people is a relatively new phenomenon. The *concept* of gender, however, most definitely isn't.

For most of its short history, the psychosocial use of *gender* applied only to humans. Because it is based on culture rather than biology, and we had assumed that animals have little or no culture, the use of the word *gender* to refer to animals seemed inappropriate. Surely animals don't have gender, only sexes. They are male, female, or hermaphrodite; maybe we can count parthenogenetic as a sex, but that's pretty much it. Well, like so many other things that we understand about animals, this too is changing. To begin with, animals certainly *do* have culture. Their behaviors are governed not just by blind instinct but also by the social milieu in which their instincts become actions, including sexual behaviors. That's culture. This was one of the major themes of my first book, *Not So Different: Finding Human Nature in Animals.*

But the bigger reason some biologists are now applying the word *gender* to animals is because we have discovered that, for many animals, there is more than one way to be a male or a female. There are different *versions* of "male behavior" and "female behavior" that animals can adopt. Importantly, I do not just mean that males or females can choose from a tool kit of different possible mating strategies. What I mean is that there are often different *kinds* of males or females in many species, resulting from real biological differences and destining them for different kinds of behaviors. While the notion of animals having genders is relatively new, animals have always had a variety of what we now call *gendered behavior.*

Even if most zoologists are still not comfortable with the term *gender* being applied to animals, none disagree with the main point here: that animals employ highly varied strategies for sex and reproduction. Besides, there currently doesn't exist any other suitable term for the ways that animals can engage their biological sex—their maleness, femaleness, or hermaphroditism—in different ways. So, *gender* it is. Animals have different genders and some people just need to deal with it.

Lipstick on a Pig

The way I use the word *gender* here was first pioneered by the eminent evolutionary biologist and animal behaviorist Joan Roughgarden, who wrote the magnum opus on this topic, *Evolution's Rainbow*, from which I borrowed the title of the last chapter (with her permission). Through a lifetime of careful research, Professor Roughgarden has taught us that the sexual behaviors of animals do not simply come down to male behaviors and female behaviors. Instead, there can be different "kinds" of males and females that approach sex differently. These different approaches are often biologically programmed and fixed, meaning that an animal cannot simply switch among them at will. We are talking about more than just variable mating strategies.

Among humans, considering gender as purely a function of sex, mating, and reproduction is unseemly. After all, being a woman in our society is about much more than just having children. But this is because we exist in a massive social milieu we call culture, and, while animal cultures and societies are much more complex than most people realize, they are still quite a bit simpler than human cultures. Therefore, we are safe—for now, anyway—to conceptualize animal genders as patterns of behavior that are either sexualized or tied to reproduction.

To a certain extent, different animal genders come about because of sexual competition. As much as I will emphasize the cooperative nature of animal communities and families throughout this book, we must also admit that animal populations are rife with intense competition. While some forms of competition are overt and conspicuous, others are more subtle. One of the greatest insights that Charles Darwin had is that every single species overproduces offspring, setting up a competitive struggle for food, other resources, and reproduction. When large numbers of animals find themselves going after the same food, for example, most of the animals simply try to up their game and win the competition. Others, however, may try to get around the competition by exploring different food options and attempting to make a living on a resource that the others ignore. If they can make use of an underutilized resource, even if it's not ideal, they can avoid the intense competition. Competition, therefore, is a major driver of ecological experimentation and innovation.

One way that competition plays out in nature is called *frequency-dependent selection*, which holds that some features are advantageous purely because they are rare. This phenomenon was first discovered in nonsexual contexts. For example, imagine a tapeworm that lives as a parasite inside our intestines. There are hundreds of sites at which the parasite can latch on to in order to feed, but not all of those sites are equally good for getting a good grip, giving access to blood vessels, etc. There are "prime" spots and "subprime" spots. However, if all of the parasites go after only the prime spots, this leaves the subprime spots unoccupied. Therefore, some tapeworms may evolve to pursue the subprime spots.[1] Those sites may not be as rich or easy to exploit, but the lack of competition makes up for that, and these pioneers will be more successful than those who are left fighting over the limited prime spots. As their offspring multiply, the competition may even switch back the other way, setting up clashes for subprime spots while the prime spots become less crowded. And so on.

Therefore, frequency-dependent selection often leads to repeating cycles in which the relative advantage of a certain strategy is greatest when it is new and unique, but diminishes as folks jump on the bandwagon. This is an excellent recipe for diversity and innovation because sometimes just being different from the herd is an advantage in itself. There are not many universal truths when it comes to biology, but one of them is that living things are diversity-generating machines.

Why would sexual behaviors be any different? For example, if the males of a given species all try to win a female mate in the same way, some will win and others will lose. For the losers, why not try something different? There's really nothing to lose and potentially everything to gain. Sometimes, just doing something different from the crowd gives one an advantage. Let's explore how this truth plays out in terms of gender.

One of the most illustrative models of multiple animal genders is the one first offered by Roughgarden herself. Bluegill sunfish (*Lepomis macrochirus*) are among the most abundant fish species in freshwater lakes and streams in North America. Anyone who, like me, has spent time fishing in lakes, rivers, and creeks in the United States east of the Rocky Mountains has probably caught a lot of these fish and thrown them back in pursuit of something tastier.

No bigger than an outstretched human hand, sunfish have a faint green-yellow coloring with lots of dark spots all over and a distinctive black spot just behind the gills, casually called an ear. They may also have yellow or pale orange coloring on their underbelly or dark vertical stripes along their body, and there is a great deal of regional variation. Many anglers assume that bluegill and sunfish are separate species of fish altogether, and while a certain area may indeed harbor multiple species of sunfish, most often what we see is sex- and gender-based variation within the same species. With a little practice, males and females are generally very easy to distinguish within the regional variation patterns.

In general, males are larger and grow more quickly than females. They also perform all of the important work when it comes to reproduction.[2] Sunfish spawn in nests that are built, maintained, and protected by the males. The female contribution to parenting is limited to a single day each year when she shares her eggs with the males of her choice, and she is quite choosy indeed. In order to convince a female to spawn with him, a male must build an impressive nest in a desirable location and convince her that he will protect the eggs until they hatch. This takes a couple of weeks of work beforehand by the male as he crafts the best possible nest, with weeks of work *after* spawning to protect the developing eggs. These males are superdads. The female investment in the offspring, however, is limited to squirting eggs, and then she's gone forever. It's a species of deadbeat moms.

The males come in three distinct varieties.[3] The males that do all of the nest-building are the largest ones, and they exhibit the most distinct coloration. They have the lightest pale-green coloring, with only faint black stripes (if at all), and they exhibit the striking flashes of orange or yellow in their underbelly, below and behind their gill flaps. Some regional variants even have flashes of orange all over their bodies—that's the distinctive sunfish coloring. Let's call these the large males. Females, on the other hand, are a bit smaller, much darker in their background green coloring, and they have thick black vertical stripes along their sides, sometimes called bars.

Ahead of spawning day, the large males gather together to build their nests in a huge spawning burrow, but they do not school together, like females do. In fact, these males are territorial and generally aggressive toward one another. The clustering of the nests seems to be driven by the "safety in numbers" principle that, together, a large group of males can more effectively defend their nests against egg predators. This is a good example of how competition and cooperation can closely coexist among members of a species. The "best" spots in the spawning burrow are toward the center, farthest from danger. Therefore, females strongly favor the largest and most central nests, so the nest-building period sees a lot of skirmishes over territory among the males.

After a couple of weeks of attempting to build and guard the perfect nest, barely eating or sleeping, the large males are ready to impress the females. When the school of females arrives at the spawning burrow, they survey the landscape and begin to inspect individual males and the nests they've built. The males attempt to physically coerce them, but females will tolerate only so much bullying before they move on. When she is satisfied with the quality of a male and his nest, a female submits to courtship with him. They wiggle side by side for a little bit, then turn at an angle to rub their genital pores together briefly before each squirts their gametes into the nest. Both of them quiver as they do this, reminiscent of an orgasm, and the whole thing lasts a little less than a minute. This process repeats thousands of times during this day-long sex romp. The males continue to skirmish with each other, occasionally trespassing into neighboring nests and attempting to squirt sperm onto the eggs there. While a given male will fertilize most of the eggs in his nest, around 10 percent of them will be fertilized by trespassing neighbors.

Meanwhile, another type of male lurks about.[4] Less than half the size of the big daddies, these males are fast and skittish. They have pale green-gray coloring, like the big males, but without the flashes of orange or yellow. They hang out at the periphery of the spawning burrow and hide the best they can in surrounding vegetation, rocks, or whatever concealment is available. At spawning time, these sneaky little guys dart in and out of the nests, squirting their sperm on any fresh eggs they can find. Let's call these the small males. Of course, the large males aggressively chase out the small ones, but they are distracted by the threat of trespassing by other large males and they also must pay attention to, you know, the females.

Besides their speed, small size, and camouflage coloring, the small males have another advantage in their quest for paternity: big balls. The testes of these little guys are around 5 percent of their body weight and take up most of the space of their abdominal cavities, the equivalent of a 200-pound human with 10-pound testicles. What these little guys lack in proximity to the freshly laid eggs, they make up for in sperm count. Small males thus release a thick cloud of sperm as they sprint through the nests. Because eggs can survive for about an hour unfertilized, the sneakiness of the small males occasionally pays off and they are rewarded with paternity of around 5 percent of the hatchlings each year.

Spawning is a very stressful activity for the large males. All they want to do is get laid but they have encroaching males, big and small, on all sides. They must be vigilant about the males and attentive to the ladies, while trying to get their sperm delivered to the nest with good aim and timing. Their testes are nowhere near the size of the small males'—less than 1 percent of their body weight—because their strategy is one of control and aggression, rather than shooting massive loads.*

---

* These are still large testes compared to humans because thousands of matings require a lot of sperm even without squirting it around willy-nilly like the small males do.

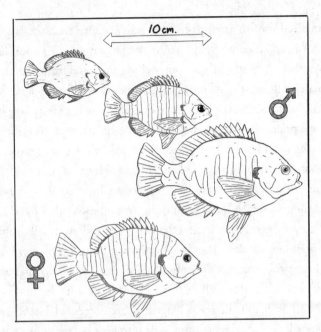

Females and three genders of male bluegill sunfish

Scientists and anglers have long mistaken the small males to be simply juveniles that were still growing up. It was assumed that the sneaky male was just immature and possibly serving an apprenticeship program in which they observed the adult male behavior in preparation for their turn in the big dance, while trying to sneak some paternity while they were at it. This view is incorrect. While they do tend to be younger, these small males do not grow up to become large nest-building males. The big guys are *born* to be big guys and must grow and develop for seven or eight years before they are ready for prime time. Importantly, they are *not* sexually mature until that point. But the small males, who become sexually mature in just three years, do continue to grow up and eventually become something else: medium-sized males.

The dance between the females and the large males is not the only court-ship that takes place. Beginning a day or two before and continuing through spawning day, a third type of male, medium in size, will approach the nests.[5] These medium males have dark bodies with dark black stripes, like the females, and without the orange or yellow flashes. They are usually mistaken for females by inexperienced humans. Instead of building their own nests, the medium-sized sunfish attempt to form partnerships with the large males. In other words, the medium males court the large ones. This courtship is a bit different, though. Rather than approaching the nests from the same level as the females will do, the medium males float down from above. When this happens, the nest-building large male doesn't chase him away, like he would a small male or another large male, and he doesn't attempt to corral and coerce him, like he would a female. Instead, he is more passive, more tolerant.

The large and medium males have their own form of courtship. It is always a one-on-one affair and is similar to the courtship that the large males have with females. There is a lot of swimming around each other for visual inspection, followed by rubbing their sides together, and eventually rubbing their genital openings as well. They even quiver. However, no sperm is released. Sperm is just too precious a commodity to waste when no eggs are present, especially considering that these fish spawn for just one day out of the entire year. Not all of these male-male courtships are successful, but when they are, the large male accepts the medium male into his nest area while they wait together as a team for the females—that is, if there isn't already a female present. Together, they attempt to woo the ladies and repel other dudes.

Whether a female is already present or joins the pair later, the spawning that takes place when a large male has accepted a medium male partner is a three-way. The females don't mind at all that a third party will be joining the fun! They all inspect each other, and the female pays a lot of attention to the size and quality of the nest. During the mating itself, the medium male is in the middle and the large male controls most of the action, although gently because females will up and leave if the large males are too aggressive. If all goes well, they rub each other on the side, and then do a sort of three-way genital rubbing, while quivering, and eventually they all release their gametes into the nest.

What is going on here? Why would the large male accept this arrangement? By bringing an additional male to his nest, he has to split the paternity, so what is in it for him? Let's look a little closer.

The largest males with the best nests tend not to accept partnership with the medium males. They already have a great position and they are happy to go it alone, stressful though that may be. For otherwise less desirable large males, the partnership with a medium male offers an alternative way to attract females. When a large and medium male tend a nest together, their reproductive interests are aligned. They're a team. And the females recognize this. Given two nests of equal quality, females actually prefer to mate with a partnered duo over a male who is working alone. For one thing, the partnered ones are less coercive. Something about pairing with a medium male pacifies the large male. I imagine that it is way more stressful doing all of this work alone, and none of us are at our nicest when we're stressed. Second, the females must know that two males are better than one when it comes to defending eggs against interlopers and predators. Eggs are precious, and the female wants her genetic posterity to be well cared for. A strong male with a large, central nest is a good investment, but so is a lesser male with a loyal helper.

For sunfish, spawning day is a blur of courtship, chasing, aggression, partnership, and, of course, sex. The large males end up with most of the paternity, around 80 percent, which is only fair since they do most of the hard work. The medium males end up with about 15 percent of the paternity by helping out with around a third of the nests. Not too shabby. This leaves just 5 percent for the smallest males. Sneaking is not a terribly successful strategy, but it does occasionally work. The smallest males are really just biding their time anyway. They will only be that way for a year, sometimes two, before they are fully grown and become the medium males.

That's right, the small males grow up to become medium males. They are sort of like sex-switching hermaphrodites except that they don't switch sex, they switch *gender*. So, genetically speaking, there are two body types among sunfish males: the large type and the small-medium type. The genetics are simple. Large males father large males and small-medium males father small-medium males. The demographics, however, are interesting because even though the large males end up fathering around 80 percent of each year's hatchlings, they comprise only about 15 percent of the adult male population. Where are all the rest of them?

There are at least three reasons why the large body type is the most common among hatchlings but the least common among adults. First, the large males must survive for *eight years* before they are reproductively capable. Small and medium males must wait only three and four years, respectively. With mortality occurring more or less steadily throughout the lifespan of these fish, waiting twice as long to reach maturity means a lot fewer get there. Second, the mutual aggression among large males can sometimes be deadly. Fighting each other to the death is something that males do in many species (and it is rare among females). And third, the prespawn weeks of nest-building and postspawn weeks of egg-guarding are very hard on the big males. It's a general rule in the animal kingdom that parental care often comes at considerable cost.* We human parents can relate.

By now you may be wondering why I chose to spend so much time giving you an up-close look at the reproductive behaviors of this one species of fish. The thing to keep in mind is that this is not some freak case. We know all this about sunfish only because some scientists have taken the time to study them carefully. The general trend in animal behavior is that every species that we observe thoroughly is way more complicated than it appears at first blush. I chose to discuss sunfish not because they are special but because they are not. They exemplify themes that run throughout the animal kingdom, and therefore throughout this book.

One lesson from the sunfish is that cooperation can exist right alongside intense competition. In fact, I might even go so far as to say that competition among animals often leads to cooperation as a way to "win." As the saying goes, if you can't beat 'em, join 'em. More accurately, we might say that the best way to beat someone is to cooperate with someone else. Reproduction is especially fertile ground (if you'll pardon the pun) for the emergence of cooperative behaviors because reproduction cannot exist without at least some baseline level of cooperation between two animals whose interests are aligned, if briefly. We tend to think of cooperation as existing only between fathers and mothers, but if two fathers can enhance their own reproductive potential by working together, why wouldn't they?

Another theme that we see in sunfish is the much greater competition among males than among females. Although hardly universal, we see this often in the animal kingdom, while the opposite situation is more rare. This

---

* In birds, for example, it has been known since the 1950s that there is a direct correlation between clutch size and mortality for many bird species. For each egg laid, the chance of surviving the year drops by a consistent percentage for the caregivers. This places the desire to have more offspring in perpetual tension with the ability to successfully raise those offspring to self-sufficiency.

makes sense when you consider the factors that determine population size and growth. Demographers of both humans and animals have long known that population dynamics depend strictly on the number of females, while the number of males is largely irrelevant. For example, cattle farmers need only one bull for hundreds of cows.

To offer a human example, the black plague devastated population growth in Europe, while World Wars I and II had only a modest effect, despite the fact that a substantial percentage of the population was killed in the successive conflicts. This is because *Yersinia pestis* infects people indiscriminately while casualties from the world wars were borne much more heavily by men. Males are not limiting for population growth so, as individuals, they are expendable. The expendability of males leads to conflict and competition with one another. In sunfish, the large males are territorial and mutually aggressive, while the females school together peacefully.

The result of that conflict leads to a third theme in sunfish that we see in many other kinds of animals as well: reproductive diversity in males. As explained above, one result of the conflict can be that individuals are forced to explore other ways to compete. The small sunfish males exhibit a particular pattern that we see in males throughout the animal kingdom: they are sneaky fuckers.* Given the intense and often deadly competition among large males, it seems only natural that sneakiness would emerge as a viable strategy, especially in fish, when mating simply involves squirting sperm in the general vicinity of eggs.

Sneaking appears in many kinds of vertebrates, including mammals.[6] In many species of deer, for example, while big ten-point bucks wrestle each other with their enormous antlers, vying for the exclusive privilege of mating with the does, smaller males without antlers often sneak up to the viewing party and attempt to mount the females as they watch the fight. The skittish does often resist and run away, but not always.

---

* *Sneaky fucker* is indeed the term used by many biologists to describe this reproductive strategy, although it is often rendered more politely in written form as *sneaky mater* or simply *sneaker*. While many attribute this term to one of the founders of modern evolutionary biology, John Maynard Smith, the earliest written use of the term I could find was by Jeremy Cherfas in 1977, followed by Richard Dawkins in 1978. In both cases, the term is used as though it was already widely known, suggesting that it had been in use before that informally.

The medium-sized sunfish males represent another alternative mating strategy that we see quite commonly: that of the cooperator. This strategy involves forming partnerships with larger, more dominant males in the hopes of gaining reproductive access. How this plays out will be highly variable based on the behavioral ecology of each species, but the common theme is that both parties can boost their own reproductive potential by working together. Sometimes the cooperation is in attracting females, as we see in sunfish, but it could also be in forming competitive alliances against other males. No matter how the beta male helps the alpha, the reward is the same: a share of paternity.

Not surprisingly, the paternity sharing is never equal nor are the partnerships universal in most species. Often, alpha males seem to be hard-wired to shoot for the stars and seek solitary reproductive dominance, like the proverbial "king of the jungle." But, speaking of lions, it is worth noting that many lion prides are led not by a single alpha male but by a coalition of males with the alpha controlling his subordinates and granting access to the females based on his perception of their helpfulness.[7] (To nurture their partnership, male lions frequently have sex with each other, reaching climax and orgasm, but that's a story for another chapter.* The next one, in fact.)

In the case of lions, the alphas and betas aren't really different genders because all of the betas would really like to be the alpha and they often try to be. But with sunfish, the three different male types are more than just strategies that a male can choose to employ and experiment with. The large type of male is born that way. The small type of male is born that way and then later further matures into a medium-type male. There is no "choice" or experimentation. It is biologically programmed and genetically determined. This is why we say that sunfish have four genders, one gender of female and three genders of male. Each has its own advantages and disadvantages in certain contexts, and each participates in the success and evolution of the species. This rainbow of reproductive diversity is part and parcel of what it means to be a sunfish.

You may have noticed that I have a lot more to say about gender diversity in males than in females. For many years, I thought that this was just how things were, that the sex that makes the small gamete experiences intense

---

* Lone-male lion families will also be discussed in chapter 4.

pressure to diversify its reproductive strategies, while the sex that makes the large gamete mostly sits back and chooses mates. That's what all the research articles say. But, the book *Bitch: On the Female of the Species* by Lucy Cooke opened my eyes to the fact that this *apparent* difference between the sexes is likely the result of patriarchal biases embedded in the biological sciences, rather than a true difference between males and females.[8] The reason there is more research on male behavioral diversity is that biologists have spent more time dissecting male behavior than female behavior.

The first truly scientific (western) animal behaviorists were called *naturalists* and were mostly nineteenth-century gentleman-scholars. Their Victorian prejudices infected their otherwise diligent and insightful study of the natural world. In *The Descent of Man* (because of course "man" means "human beings"), Darwin declared that, "The males of almost all animals have stronger passions than the females . . . The female, on the other hand . . . is less eager."[9] This both reflected and perpetuated the view that males are the drivers of behavior, while females are merely passengers. And this view was not just applied toward the pursuit of sexual intercourse, but to most kinds of behavior (except child-rearing, of course). It is my hope that books like *Bitch* will raise our collective consciousness as a new generation of animal behaviorists prepares to take over the reins of the field, with women approaching equal representation. I suspect that we are only just beginning to appreciate the full picture of gender diversity.

If You Can't Join 'Em, Beat 'Em

Having described the concept of gender in animals through the example of sunfish, we can explore more gender diversity throughout the animal kingdom, inching our discussion progressively closer to our own species. After all, sunfish are members of a group of vertebrates called *Osteichthyes*—the bony fishes, which have been evolving separately from our lineage for around 400 million years. They are not close relatives, so we should avoid trying to draw straight lines between their behaviors and ours. The lesson from sunfish is that the sex of an animal does not strictly determine its sexual behaviors. There's always more than meets the eye.

Before we leave seafood behind, there is another familiar fish that exhibits some instructive gender diversity as well. Both the Pacific and Atlantic species of salmon have similar lifestyles in that they are hatched in freshwater streams, migrate to the open ocean where they spend most of their lives, then return to their natal fresh water, swimming upstream and jumping waterfalls, to spawn and die. They have something else in common as well: both have two genders of males.

In the Pacific, the females and the hooknose male salmon (*Oncorhynchus*) spend at least three years in the open ocean, eating and growing, before swimming upstream for spawning.[10] But there are also other males called jacks that spend only two years out at sea before heading inland. While the hooknoses are larger and aggressively guard their mated female and their nest, the jacks are smaller and faster. They are the sneaky fuckers of this species, darting through the spawning grounds spreading sperm where they can.

Head-to-head, the large hooknoses easily outcompete the jacks, but the jacks are more numerous and appear to have faster, higher-quality sperm, resulting in a delicate balance of the two forms. A similar situation exists in Atlantic salmon (*Salmo salar*) except that the sneaky jacks never venture into the open ocean at all.[11] They simply stay in the streams and begin their sneaky mating strategy as soon as they are reproductively able. The larger hooknoses come back from the ocean when they are at least three years old; the largest of these can be over ten years old.

Like sunfish, male salmon are an excellent lesson in how a species can develop diverse mating strategies that coexist in competitive tension. There is great value in this because it adds to the diversity of the survival tool kit of the species. For example, if the ocean were to become more hostile to salmon for any reason, the jacks would have an advantage and the species could easily adapt to a more freshwater lifestyle. Same with lifespan. In general, during times of environmental stability, longer lifespans are advantageous, but during periods of abrupt environmental change, shorter generation times are often key for adapting to the change and surviving. By having multiple genders of males with different lifestyles and lifespans, a species has built-in diversity that it can tap into when needed. That's enough about fish for now.

White-throated sparrows (*Zonotrichia albicollis*) are among the most common backyard songbirds in New England and Eastern Canada. Unlike most songbirds, the females are just as musical as the males. Male-female courtship involves singing by both parties and, once paired, the couples work together to aggressively attain and defend a territory, build their nest together, sit on the eggs, hunt for food to share, and cooperatively co-parent their offspring. The division of labor is more or less equal, which is not uncommon in birds.

Interestingly, both males and females are strikingly colored and each comes in two varieties, called *morphs*, that involve different striping patterns on their heads.[12] For simplicity, most birders refer to them as the white-striped and tan-striped morphs, even through the differences are more detailed than that. In both sexes, the white-striped birds are larger, more aggressive, more musical, and all around better at getting and keeping a territory. Yet, almost all of the white-striped birds, male or female, pair up with a tan-striped bird, rather than another white-striped bird, even though the latter is inferior in their singing, aggression, and territoriality. Why would a large, aggressive female specifically seek out a smaller, less territorial male (and vice versa)?

It turns out that what tan-striped birds lack in aggression and territoriality, they make up for in parenting skills.[13] They are more successful in hatching eggs and in helping the hatchlings grow up and thrive. This is quite a remarkable dichotomy. In birds, there is usually either a broad sex-based dichotomy or there is no dichotomy at all. Of course, in any species, some individuals are more aggressive than others and some are better parents than others. And some species employ shared or communal alloparenting, and so on: this is a case in which we have distinct morphs with different parenting roles, but those morphs are entirely disconnected from biological sex.

Therefore, we might say that white-throated sparrows have four genders: white-striped females, white-striped males, tan-striped females, and tan-striped males. Or, since the pairings are almost exclusively heterosexual *and* opposite-striped, it might make more sense to say that there are two sexes and two genders, and that the two categories are independent of one another. However we arrange the semantics, these birds have clearly achieved a complete ceasefire in the battle of the sexes while still retaining distinct "roles" for the key behaviors related to reproduction: the establishment and defense of a nest territory and the rearing of the young. How progressive!

Ruffs (*Calidris pugnax*) are a species of small migratory sandpiper birds found all over Europe and Asia. The females have dull brown and white coloration, but the males have a bright tuft of feathers surrounding their necks that they can puff out like the Elizabethan neck adornment that lent these birds their name. Most male ruffs are aggressive and territorial, and yet, like sunfish, they group together to build and defend their nests in clusters called *leks*. In addition to the puffed-out neck feathers, male displays also involve elaborate calisthenics including jumping, crouching, twisting, head bouncing, lunging, wing fluttering, and so on. It looks like dancing done by someone with absolutely no skill in dancing (like myself). The displays are not directed at females but rather at other males and are used, along with general aggression, to establish a dominance hierarchy within the lek, which determines nest placement and access to the choicest females. When the females arrive at the lek, they begin to mate with the males according to the preestablished dominance order.

But that's only *most* males. Around 15 percent of male ruffs are a little different. They don't bother making or defending a nest, they don't do the silly dance, and their neck plumage is underwhelming.[14] And yet, there they are at the lek, tolerated by the other males (more than they tolerate each other) and competing for access to females, independent of the pecking order observed by the other males. They don't even stay at the leks. After the breeding period is over, they vanish, leaving the work of caring for eggs and hatchlings to the females and the other males. They are sneakers of a sort, but they do so brazenly, in full view. Why don't the other males simply chase them out?

It turns out that flocks of female ruffs are attracted to leks according to how many males are there. More males means more mating opportunities, more genetic diversity, etc., but simply having more of the "main" males isn't always an option because a given lek can support only so many nests before it really isn't a good lek anymore. Too many birds in residence means that the waste situation will become unsanitary and it is more likely to attract large predators like wolves and bears against whom the ruffs have no hope of defending the lek. So the extra so-called "satellite" males are there to artificially and temporarily puff up the numbers of males and therefore attract more females, without contributing to overpopulation among the permanent residents.

---

* There are actually two distinct subtypes of these "different" males, which are easy to distinguish from each other visibly but otherwise less understood.

The reason ruffs have evolved these particular types of males when other birds haven't (that we know of) is because they practice a lifestyle that is not common among birds: polyandry. Most female ruffs mate with several males, which drives the male-male competition seen in the display behaviors, but it also drives something called sperm competition and, accordingly, male ruffs have very large testes for their body size. In fact, ruffs exhibit the greatest degree of polyandry of any known bird species (measured by percent of females that pair-bond with two or more males in a given year) and they also have the largest testes relative to body size. This is not a coincidence.

In ruffs, the satellite male gender has emerged to help the main males attract more females. They are cooperators, of a sort. In return they get a small slice of the paternity without having to do any nest-building, display competition, defend eggs, or care for hatchlings. It's not a bad deal! Once again, we see that a unique mating style by females—in this case, polyandry—drives the evolution of gender variety in males.

Fascinatingly, ruffs represent one of the few examples in which scientists have successfully unraveled the fine molecular details that underpin their unique gender diversity. In 2016, two separate research groups made the startling discovery that the ruff genome contains a so-called *supergene*, a group of genes that are clustered so tightly together that they are inherited as though they were a single gene.[15] This supergene is responsible for the satellite male gender and encodes both the behavioral peculiarities and the distinct color pattern of these males. The existence of this supergene helps explain how the physical traits and the behavioral traits are so closely linked together genetically in these birds.

Subsequent studies published in 2023 confirmed that the ruff supergene is the result of an inversion—a type of mutation in which part of a chromosome is removed and then put in place backward—and that this inversion appeared very recently, within the past 70,000 years.[16] Thus, ruffs are a case in which random molecular tinkering resulted in a variant that was successful, but only in partnership with others. The supergene that encodes the satellite behaviors will never fully take over in the ruff population because satellite males cannot compete against the "independent" morphs. But they *can* exist alongside them, and when they do, both types of males benefit. Satellite males are therefore a type of helper male, and over the past 70,000 years of natural selection, this morph has settled into the ruff gene pool at about 15 percent frequency.

Priscilla, Queen of the Desert

While "gender diversity" is not a common term in the biology literature (yet), there are hundreds of studies on what is called *female mimicry* in various animals. This is the idea that, in some species, males have evolved to use trickery against both females and other males to gain some kind of reproductive edge. Scientists believe that these tricksters use mimicry to avoid territorial disputes with males and/or to sneak up on unsuspecting females. In other words, this phenomenon is the animal equivalent of cross-dressing, and some examples below show how this can be an effective strategy in certain circumstances.

Males in a bird species called the Western Marsh Harrier are thought to use cross-dressing to compete for reproductive access. These birds are raptors, meaning that they are predators like hawks or eagles, and part of their scientific name—*Circus aeruginosus*—comes from the familiar way in which they *circle* their prey while hunting from the air. Like most raptors, these birds live in low population densities and are aggressively territorial. Like sunfish, the males build nests to impress and attract females, but, unlike sunfish, they will not tolerate another male nearby. They fiercely defend the area around their nest with a radius of almost a kilometer and will dive-bomb any male that encroaches in their space.

Except for the female-mimicking ones. The majority of males are gray with yellow eyes, while females have a more brownish color, with striking white feathers on the top of their heads. The white feathers on top help females distinguish themselves from males so that they don't accidentally get attacked from above. Some males, however, exhibit the white feathers also and they do so in order to avoid territorial disputes with the other males. Their camouflage allows them to move through the various territories unbothered by males as they search for females with which to mate. Apparently, other males are easily fooled and the female mimicry works, as these "transvestite raptors" (as *New Scientist* magazine called them) obtain a rather respectable percentage of paternity.[17]

The discovery of female mimicry in this species made a splash in 2011 because this phenomenon, while more common in insects, fish, and reptiles, had not been observed previously in any bird of prey. But the thing is. . . . it's bullshit.

Despite the claims of the authors of that study, I do not believe that male western marsh harriers camouflage themselves as females. I base my skepticism solely on data in previous articles from this very same research team and what we know about bird physiology. Let's keep in mind that birds of prey have *the best visual system of any living thing on the planet*, hence the expression "eagle eyes." In terms of visual acuity at a distance, raptor eyes are about *five times* more powerful than ours, and yet they are easily fooled by a few white feathers? These supposed female mimics are the same size as the other males, which makes them about 30 percent smaller than the females. (That's twice the average size difference in adult human males and females.) Also, the "female mimics" have the distinctive yellow eyes that the other males have. If this is female mimicry, these are the worst drag queens ever.

In fact, I find most descriptions of so-called female mimicry to be highly suspect. It is not my intention to impugn the integrity of the many scientists that study so-called female mimicry among animals, nor do I wish to question their intelligence or insult their life's work, as many of them are more accomplished scientists than I. Nevertheless, I am deeply skeptical that "female mimicry" is what is going on in many of these cases, and in some, I am utterly convinced that it is not.* I will concede that it is certainly possible that the phenomenon is real and that it exists in some species of insects, spiders, fish, and maybe others with rudimentary nervous systems. However, I think the few examples of female mimicry in birds and mammals that have been studied in detail are better interpreted under the theoretical framework of gender diversity rather than deception. Admittedly, this framework is still only nibbling around the edges of the scientific mainstream, but a shift may nearly be upon us.

---

* Once again, it is Joan Roughgarden whose writing has opened many eyes on this topic, including my own. I am mostly parroting—or shall we say "mimicking"—her criticism of the concept of female mimicry. Her book *Evolution's Rainbow* critiques this concept in more detail and with different examples.

Perhaps the alternate plumage pattern of some male marsh harriers is not meant to fool anyone into thinking that they are females but rather to signal—honestly—that they are not a regular territorial male and harbor no wish to challenge the tough guys. The white feathers on the top of their heads are like a flag saying "I come in peace." Why would the other males accept the olive branch and willingly leave them alone, even knowing they are males? It turns out there is another reason that these males are special besides their androgynous coloring: they are experts at *mobbing*, a collective behavior for rooting out nest predators.[18]

Nest predators for harrier eggs include small rodents, dogs, wolves, foxes, skunks, badgers, and even other birds. Accordingly, the birds have evolved the elaborate anti-egg predator defense called mobbing, which is exactly what it sounds like. When a potential predator is spotted, a bird will issue a specialized call, and nearby birds band together to scare off the would-be egg thief. This may sound confusing, since I said earlier that the males are territorial and don't tolerate the presence of another male within hundreds of meters, let alone act together with them in a roving mob. But there's no conflict here because *those* males don't bother to participate in the mob. Instead, the mobs are formed of females and the special type of male, the so-called female mimics. In fact, the scientists that study these birds consider their mobbing behavior part of their female mimicry.

Or maybe this is just another male gender. Rather than fighting it out with the other males, these guys have figured out another way to succeed: by cooperating. Sound familiar? In this case, instead of cooperating with other males like the medium-sized sunfish males do, they cooperate with females in chasing out nest predators. Their white feathers signal to the territorial males that they are no threat, while signaling to the females that they are on the same team. When it comes to mobbing, the regular males aren't exactly deadbeats because someone has to stay to defend the nest. But the mob is bolstered by the addition of these "alternate" males. They are cooperators. And like cooperators in other species, they add some typically female markings to their otherwise male bodies so that the other males can recognize them and treat them differently. This seems much more plausible than the males being fooled.

Furthermore, the demographics of marsh harriers provide a surplus of unpaired males. The males who woo females form pair-bonds with them, but they do so in a polygynous arrangement. A given nest includes a single male and one, two, or even three females. This leaves a lot of extra males. One reproductive strategy is to just fight it out and struggle to build and defend a nest. However, when competition is intense, evolution rewards creativity. The nonnesting males have found a way to make themselves useful by helping to defend the nests of their fellow conspecifics. Of course, they don't do this just to be nice. They expect—and obtain—some paternity for their troubles via extra-pair copulations (EPCs).* This is an alternative mating strategy, but it only works as an adjunct to the main strategy of nest-building. Further still, we would expect to observe such alternative strategies only in species that have a lot of unpaired males. Something like 90 percent of migratory birds form "seasonally monogamous" pair-bonds, and that's probably why we don't see alternative types of males very often in birds. But in one of the few species that practice polygyny, the western marsh harrier, we observe an ecologically important alternative male morph. Once again, I do not believe that this is a coincidence.

Rather than female mimicry, why not consider this another male gender? To summarize the evidence, the mimicry is incomplete with regards to body size and eye color; the birds have excellent vision; the supposed mimics employ a strategy that is useful and important; and the most convincing piece of evidence of all is that the cooperator males are not few and far between. They comprise around *40 percent of the male population*. It is common sense that, in general, camouflage can work for an individual here and there, but not for masses of animals. The more common a particular variant is in the population, the more familiar the others are with it. It seems to me that the framework that best fits the evidence is that western marsh harriers have two genders of males, nesters and mobbers. Both serve an important function and both get rewarded with paternity when they do.

Let's consider another supposed female mimic. The European pied flycatcher (*Ficedula hypoleuca*) was the first bird species in which female mimicry was proposed.[19] This is another interesting bird for our discussion because they do not practice the seasonal monogamy that is typical of most

---

* Extra-pair copulation is a very common mating strategy in birds, which will be explored in detail in chapter 4.

migratory birds. Instead, the males of this species are thought to be deceptive polygynists and are polyterritorial. In the spring, males keep very busy securing and defending a nesting site, usually in an oak tree, building the nest and courting a female mate for the season. Once this is done and the female has laid her eggs, the male attempts to repeat the cycle with a second female—or, I should say, a *secondary* female. She is considered secondary because, once a second clutch of eggs have been laid, the male returns to his primary mate and resumes his share of the typical child-rearing duties, bringing food to the chicks and so forth. The secondary mate is on her own and, not surprisingly, she and her offspring are less likely to successfully survive the year than is the primary mate.

It's pretty obvious why the male flycatcher does this. While he continues to invest most of his energy on the offspring he has with his primary mate, there is little cost and considerable potential gain by spreading his seed, so to speak, in the hopes that some of the secondary offspring will defy the odds and be successful. Yes, the male must secure and defend a second territory, but not for long. Once he convinces a female to join him and he does the deed, he is gone. Small investment, high risk, high potential reward.

Interestingly, the polygynist males always make it a point to keep their two lives—that is, their two *wives*—far apart. They never pursue adjacent territories. In fact, their two nests are usually located many hundreds of meters apart, up to several kilometers even. The males seem to go out of their way to keep their mates from bumping into each other. Wise, those flycatchers, because, as it turns out, females of the species are also territorial and aggressive toward one another. Female-female aggression is not common in monogamous birds, but polygyny sets up a system in which females are in competition for high-quality males. Rather than a battle between the sexes, it's a battle *within* the sexes.

The long distance at which the males keep their two nests has led researchers to consider this practice "deceptive polygyny," akin to a human man who has a "side family" that is completely hidden from his wife and legitimate children. This explanation doesn't hold much water for me either. Yes, the males keep their mates far apart, but, rather than an attempt to deceive them, it seems to me that the simpler explanation is that he is simply preventing them from fighting. He has a shared genetic legacy with both of these females, and should they enter into combat, no matter which of them wins, he loses. Best to keep them far away from each other. Our tendency to routinely invoke deception as an explanation for bird behavior says more about our species than theirs.

Okay, so what about the female mimics? Once again, we see a perfect situation for an additional male gender to emerge. With the large and greedy flycatcher males pursuing multiple mates, this leaves a substantial number of unpaired males in the population—ripe breeding ground for gender experimentation. Both male and female flycatchers exhibit dull coloration from birth through sexual immaturity; most often, males gain their striking coloration upon sexual maturity. However, some young males retain the dull coloring for a full year after they mature. These are the so-called female mimics, and the common explanation is that they are tricking other males in order to avoid their aggressive territoriality. By doing so, they can sneak up on mated females and attempt to inseminate them through extra-pair copulation.

The explanation of female mimicry has some problems. First of all, as I said above, females are aggressive toward other females. Female mimicry would simply trade aggression from one sex for aggression from the other, and that's worse because now the aggression would be coming from the very individuals with whom the mimic is trying to mate. Second, with so many males attempting to guard two territories, which are also distant from each other, a "sneaking" mating strategy hardly requires a disguise of any kind. Third, the paired females are sometimes receptive to the young interloping males, but even when they're not, they don't chase them out like they would another female. They simply deny them the EPC. In other words, they don't appear to be fooled into thinking these young males are actually females.

Once again, the simpler explanation is that these males represent another form that maleness can take among flycatchers. No mimicry, no deception, no suckers. But, you may protest, don't female flycatchers prefer to mate with the toughest, scrappiest males with the best nests and territory, as a way to ensure high-quality genes for their offspring? Of course they do, and that is precisely where diversity comes in. There is no one best way to be a flycatcher or anything else. Being strong and smart enough to build and defend a nest territory within a wooded landscape is a fine measure of worthiness for flycatchers *how and where they currently live*. But those same skills would be absolutely worthless for a desert-dwelling turkey vulture, for example. So, while the sexual preferences among female flycatchers strongly favor the skills that appear to be a good proxy measure for overall strength, intelligence, and vigor, there is also some diversity in their sexual preferences. The young sneaky interlopers occasionally get some play.

This begs the question: What could these young supposed female mimics possibly offer the females in terms of genetic value, especially for those females who already have a strong mate with a high-quality territory? First, remember that just being different can be attractive. To a certain degree, animals seem to be drawn to oddballs that are otherwise healthy, strong, and successful. Mixed in with the more obvious and practical sexual attractions, there is also an attraction that stems from the inherent wisdom that diversity itself is desirable.

Somehow, the wisdom of diversity appears to have been largely missed by the biologists who study flycatchers. Since the time that the supposed female mimics were first discovered in the 1980s, they have remained committed to the deception narrative despite the obvious problems with it. In one study, the scientists were so incredulous that they labeled the female behavior "imperfect mate choice," as if the females must be really stupid to choose these weird males. The article—published in one of the top journals in the field of animal behavior—includes this stunning sentence right in the abstract: "The seemingly nonadaptive, imperfect female behavior suggests that the evolution of female selectivity is constrained, allowing deception by males to be successful in the evolutionary 'arms race' between the sexes."[20]

These scientists were so blinded by their bias that the only interpretation they could come up with is that male flycatchers are brilliantly deceptive and the females are hapless idiots unable to make good choices. Prejudiced by the notion that males should be strong and aggressive, anything less is seen as a sissy, a female mimic. And prejudiced by the notion that smart females should always choose the strongest, most aggressive males, they called their mate choice "imperfect." But this raises a second problem with their explanation: these males are not sissies!

As I hinted above, females usually resist the attempted EPCs from sneaky young males. That resistance is probably a form of quality control, similar to *mate contest competition*, when males fight each other for sexual access to females. Biologists believe that this is the main reason why females resist

A female hyena and her phallic clitoris

copulation with males in so many species. By placing a hurdle in the way, a strenuous physical challenge—resisting sex—the females ensure that only strong and healthy males will father their offspring. With the main type of flycatcher males, females use their territory-defending and nest-building behaviors as their proxy indicator of robust genes,[21] so with the young sneakers, they need a different indicator. So they make these males overpower them before they relent to sex. Here we see again that this example of supposed female mimicry has a far simpler explanation: gender diversity. In science, the simplest explanation is usually the correct one.

Just so we don't leave females out of the mimicry game altogether,* some claim that female spotted hyenas exhibit "male mimicry." Here it is hard to keep my criticism respectful because the claim is that *all of the females* are male mimics. All of them. Just because they are aggressive, dominant, have high

---

* As will be discussed in chapter 4, among cuckoo birds and their relatives, some females exhibit male mimicry as a form of camouflage, but this is to fool birds of *other* species, not their conspecifics.

circulating testosterone, and sport a clitoris so large that it is usually mistaken for a penis, they must be male mimics??? Well, actually, because their urethra runs through their enormous clitoris and they urinate with it, it is technically a penis. Um, okay, I'll admit that female hyena physiology is a bit odd compared to other mammals. And one last thing about that huge clitoris: . . . Females often have sex with each other by inserting one inside the other. Oh, and they also give birth through it.

Yes, yes, it all seems very strange, but rather than declare that hyenas are a whole species of males and male mimics, perhaps the simpler explanation is that this is just what spotted hyena females are like. No mimicry, no deception, just females that don't fit the typical mold. Circulating testosterone is extremely high in spotted hyenas—in both males and females—and given that average circulating testosterone correlates pretty well with intraspecies aggression, it's no surprise that hyenas, both males and females, are aggressive, competitive, and pretty nasty toward each other. The testosterone seems also to have masculinized the female body plan to a degree that we struggle to recognize it as female. It is important to remember that that is our problem, not theirs. Hyenas are a successful species. Diversity wins again.

We are now ready to consider the gendered behavior of species much more closely related to us: our fellow African apes, the gorillas and chimpanzees.

Gorilla Gender in the Mist*

Huge swathes of central Africa's rainforests remain unexplored by Western science, but the forests of Virunga National Park have been the object of intense scientific scrutiny since George Schaller and Dian Fossey began their pioneering work there in the 1950s. Since 1967, the populations of mountain gorillas in Virunga have been the subject of continuous monitoring, nu-

---

* This section originally appeared within a longer blog post at *Gender and Society* and *The Human Evolution Blog* and was co-authored by Stacy Rosenbaum of the University of Chicago and Lincoln Park Zoo. We do not speak on behalf of the Dian Fossey Gorilla Fund International, which has not reviewed or endorsed the views represented here.

merous documentaries, and the Oscar-nominated biopic *Gorillas in the Mist*, with Sigourney Weaver portraying the iconic and tragic Dian Fossey.

Much of what we know about the ecology and social behavior of gorillas stems from the constant observation of the Virunga mountain gorillas.[22] Many common primate behaviors were first discovered in this population, including *male contest competition*, when males fight for access to females over whom they maintain exclusive mating rights; *sexually selected infanticide*, when males kill other males' infants in order to bring females into heat and redirect their maternal attention on future children; and *scramble feeding competition*, reliance on food sources that are not monopolizable, which minimizes the utility of female dominance hierarchies. Just in these three examples, you can see how "gendered" gorilla life can be.

Indeed, gorillas display substantial sexual dimorphism. Specifically, the males are more than twice as large as the females and much more powerfully built. Their size and strength is not a function of their ecology: gorillas are vegetarians. Rather, it underscores their evolutionary legacy of male contest competition and polygyny. Gorillas were long thought to exist exclusively in harems, small multi-female groups led by a single powerful silverback male. Since silverbacks eat mostly leafy greens and seasonal fruit, their herculean strength and sharp canine teeth are used only for fighting each other.

Typically, young males leave their birth group when they approach maturity and begin to represent a potential threat to the alpha silverback, who is most often their father. Rather than attempt a challenge that will surely be unsuccessful, adolescent males go off on their own and experience a solitary period as they finish their growth and maturation into full, powerful adulthood. Later, they will make attempts to take over another harem by killing the silverback (and then all the juveniles) or perhaps they will try to lure females and start a group of their own. Most are not successful. For any straight men that are jealous of the harem lifestyle of gorillas, keep in mind that most males do not get the glory and instead die in a violent conflict. Polygyny is actually pretty harsh on the vast majority of males in any species that engages in it, including ours.

Beginning in the 1990s, something unexpected happened.[23] Some younger males stopped leaving their natal groups and the basic harem social structure fundamentally shifted for about a quarter of the park's mountain gorilla population. Instead of groups with one or occasionally two adult males, scientists began observing very large groups that included several adult males and females living together in relative harmony. Some groups have hosted up to nine adult males, and one group topped out at sixty-six total animals, around *four times* the normal size of a gorilla group. Over thirty years in, this trend shows no sign of reversing, hereby turning everything we thought we knew about the reproductive behaviors of mountain gorillas on its head.

The many scientists who study the Virunga mountain gorillas have yet to arrive at a convincing answer for why these animals suddenly overturned an apparently long-standing behavioral legacy.[24] Regardless, their story is a fascinating example of social flexibility in a long-lived, big-brained primate that is not that different from us. When Schaller and Fossey first observed them, it would have been virtually unimaginable to see six adult males co-existing, but this is now routine. While these males are rarely close social partners, they work together to exclude outsiders and are remarkably tolerant of each others' mating behaviors.

The mountain gorilla silverbacks even "parent" other males' infants, and their collective protective abilities mean that infants in multi-male groups are less likely to die of infanticide than infants born in a group where their father is the only male. These males seem to have realized that cooperation can be preferable to competition! Unsurprisingly, females apparently modified their social behavior in turn. Scientists have established that there are reproductive benefits for both sexes to living in multi-male groups. It should also be noted that it is entirely possible that the females participated in—or even instigated!—this social shift, rather than simply reacted to it. Given the strong role of alpha females in the day-to-day social lives of mountain gorillas, this seems more than plausible.

We simply do not know how this remarkable transition occurred, but it shows, clear as day, that behavior is flexible and adaptable, rather than strictly hard-wired. In this case, the highly gendered behaviors of mountain gorillas, around which the entire social structure was built, shifted completely and a new structure emerged without conflict or catastrophe. Even in gorillas, gender is a social construct.

An Ace in the Hole

Having spent five decades studying primate social behaviors, Professor Frans de Waal was probably the world's leading expert in chimpanzee behavior until his death in 2024, an event that saddened the entire primatology, ethology, and anthropology communities. In his seventeenth and final book, *Different*, de Waal decided to tackle the issue of gender variation in primates, especially apes and humans, and introduced us to Donna, a chimpanzee he had known and observed for many years.[25]

Donna is female in terms of gametic sex, but she never adopted typical female behaviors and never became conspicuously receptive to sex in the way that female chimpanzees typically do, with swollen external genitalia. She is larger and more robust than females typically are, though not as large as a male, and more aggressive and domineering, as male chimpanzees tend to be. Donna is certainly a gender-bender—and that is not a common occurrence among chimpanzees, so she doesn't seem to represent an additional female gender. In addition, she shows no interest in sex whatsoever and has not reproduced to date. To be considered a gender, there would need to be some kind of reproductive strategy to her way of being, and we don't see any signs of that.

Instead, de Waal calls her asexual and gender nonbinary. Because these two labels can be described, if insufficiently, using behavior alone, he is more comfortable with those terms than in referring to her as transgender, and that makes sense to me, too. That said, it is rather easy for me to also see in Donna a form of gender diversity that, like all forms of diversity, has popped up in this chimpanzee at random. This is how variation and natural selection usually work. Variations appear at random, and then they either proliferate or disappear based on whether or not they bring advantages or disadvantages to the individuals.

In this case, Donna's variation in the way she engages gendered behavior appears not to be beneficial, since she has not reproduced and shows little interest in doing so. But it is not hard to imagine how other forms of gender variation could be a successful long-term strategy, and this chapter is filled with such examples. Importantly, each of those variations began the way that

Donna's did: randomly. Nature is constantly tinkering with our genetics and our behaviors, and gendered behaviors are no different.

It is tempting to apply the framework of animal gender diversity, as explained in this chapter, to transgender humans, but this comparison is not straightforward. For one thing, the transgender state in humans depends very much on subjective inner feelings, whereas in other animals, all we can do is describe behavior. We have no access to their private feelings and mental states.

Or do we? After all, outward behaviors are driven primarily by internal drives and emotions. We drink (a behavior) because we are thirsty (a drive). Perhaps gender identification—the urge to act like a man or woman—can be understood in the same framework. Perhaps we all have a vague drive to "act like a man," or "act like a woman," and what that precisely means depends on the time, place, and culture in which we live.

If we think of gender identity as the drive to behave in male-typical and female-typical ways, then it's not so crazy to think of Donna as a transgender chimpanzee. For the most part, she has a typical female body (her sex), but she behaves in more male-typical ways (her gender). True, this does not capture the internal experience of being transgender, but that's going to be a species-specific experience anyway. Of course Donna doesn't experience being transgender the way that humans experience it, because she doesn't experience *anything* the same way that humans do. Humans and chimpanzees are different, especially when it comes to things like our internal subjective experiences. And of course Donna is not preoccupied with the *validation* of her internal subjective feelings the way that humans are, but that doesn't mean that she doesn't have them.

Rather than arguing about what label best fits, we can simply see Donna as a unique example of the kind of gender diversity that nature, in her creative wisdom, is constantly generating. There is really no need for labels, categories, or inclusion criteria when we can just simply appreciate Donna exactly as she is.

Chapter 3
That's Gay

Humpback whales (*Megaptera novaeangliae*) are one of the greatest success stories of cooperative conservation efforts resulting in saving a species on the brink of extinction. The global population of these enormous and majestic creatures once dwindled to just a few thousand individuals in the late 1960s. In fact, the rallying cries to save these gentle giants were key to galvanizing public support for the Endangered Species Act of 1973. Due to international protections and a global ban on whaling,* humpback populations have slowly recovered. In 2021, humpbacks were officially upgraded to the status of "Least Concern" by the International Union for the Conservation of Nature (IUCN), and an estimated 150,000 whales now populate our oceans.

Although humpbacks frequently dazzle whale watchers with their acrobatic breaching, when it comes time for courtship, calving, and other reproductive behaviors, these whales become incredibly shy. Very little is known about the sex lives of humpbacks. We know that they sing. A LOT. Males have been known to sing for seven hours straight. And the songs have an incredible amount of complexity, with researchers cataloging various units, phrases, and themes, complete with familial and social relatedness and geographic variations. But we have yet to definitively conclude the function and purpose of the songs.

Humpbacks will not breed in captivity and, until late 2023, scientists had never directly observed humpbacks having sex in the wild. Finally, marine biologists were able to catch a pair of humpbacks getting busy off the coast of

---

* Not to mention the impact of *Star Trek IV: The Voyage Home*, in which the extinction of humpback whales causes a cataclysm that nearly destroys the planet!

Oahu and published their observations in *Marine Mammal Science* in February of 2024.[1] The copulating humpbacks were both male.

For centuries, a key plank in the reasoning for the persecution of gays and lesbians was that homosexual behavior is unnatural. The word *unnatural* can mean lots of things, but what is usually meant by reactionaries is that sexual activity between members of the same sex is not observed among animals in nature and therefore, by extension, isn't natural for humans either. It is worth noting that *no one* who says this actually knows or cares anything about the sex lives of animals. They do care about the sex lives of humans, for various reasons, and are trying to advance the position that sexual activity should be restricted to heterosexual couples and usually within the context of a monogamous marriage. Of course, the objection to nonsanctioned sexual activity is almost always based in religious or moral disgust, and the appeal to a supposed biological fact is purely an attempt to bolster that position without appealing to a religious doctrine.

Beginning in the Middle Ages and reaching its zenith in the nineteenth century, Western culture classified gay sex as a so-called crime against nature. Being homosexual was often referred to as having "unnatural desires" and so forth. The implication, I suppose, is that if a certain behavior isn't natural, it's okay to condemn and criminalize it. I have never heard similar arguments against using soap or wearing clothing or any of the other myriad human-specific behaviors, but despite the flimsy reasoning, this particular argument seemed to hold sway at the time. To this day, you can still come across the expression "even dogs don't do that!"

There's just one problem with this argument. Dogs most definitely "do that," as anyone who has ever owned a dog can tell you. Male and female dogs both mount and hump other dogs of both sexes. As we'll see in this chapter, this is true for basically all mammals. And animals do this with each other for a variety of reasons, and in so doing, they achieve sexual gratification. And that may even *be* the reason. Yes, animals have sex, including with members of the same sex, purely for pleasure, but, as I will show here, sexual activity often furthers a specific goal, at least broadly speaking, and this is true for all animals, including us. Yes, the pursuit of pleasure is what *drives* sexual behaviors, and so this seems *to us* to be the primary goal. But in evolutionary reasoning, pleasure is used as the incentive to drive an animal to pursue a specific goal. In more technical terms, pursuit of pleasure is merely the *proximate* goal. There are separate *ultimate* goals, among which is reproduction. Sometimes.

Traditionally, we were told that the ultimate goal of sex was procreation. No less, no more. If we view the drive to have sex as purely directed toward reproduction, then yes, of course gay sex is pointless and homosexuality is a dead end. But by this logic, oral sex would be just as pointless, as would be anal sex, masturbation, and even vaginal intercourse outside of a female's fertile period. None of these lead directly to procreation, yet they do not elicit quite the same reactionary vitriol as gay sex does. At the same time, the non-procreative nature of these other sex acts have indeed led to their derision, condemnation, and even criminalization in certain eras and jurisdictions, but none so fiercely or universally as homosexuality. Come to think of it, why has sex with a postmenopausal woman never been outlawed?

The absurdity of all this just shows how incorrectly we have viewed sex all along. As I will show in chapter 5, sexual activity advances all sorts of goals and purposes, both for the individuals involved and the community of which they are a part. The pleasure of sex is just a ruse. It's evolution's way of tricking us into aggressively pursuing sex, like a parent who uses candy to toilet train their child. The candy is all the kid cares about, but it's the toilet training that's the real goal. We'll explore the various goals of sex later on, but first we have work to do to dispel the notion that animals of the same sex don't have sex with each other.

Don't Let the Bed Bugs Jab!

It has been known for some time that male bed bugs (*Cimex*) will mount other male bed bugs and inseminate them. Bed bugs survive almost exclusively on mammal blood and can go long periods without feeding, possibly up to a year. While waiting, they will not attempt to subside on any second-choice foods. It's blood or starvation. When it comes to mating, however, they are only picky in one way: males will only mount another bug that has recently fed. Presumably, this is to ensure the best chance of success for the development of the fertilized eggs.

Their method of mating, called *traumatic insemination*, is quite strange and found in just a handful of invertebrates.[2] When a bed bug mounts another, he uses his *aedeagus* (the arthropod equivalent of a penis) to pierce his partner's abdomen and then injects his sperm right into the abdominal cavity. The sperm is then picked up by the circulation and whisked all throughout the body. If the pierced and injected bug is a female, the sperm will eventually reach the ovaries and fertilize the eggs there. If the pierced bug is a male, the sperm cells will just wander around pointlessly until they die. This mode of copulation is odd because it is very clearly injurious. At the very least, the mounted bug has a wound that is at risk for infection, but more serious damage can also result. (But this is hardly the only example of sexual behavior that results in injuries.)

Female bed bugs have no vagina or genital duct, so traumatic insemination is the only way they can reproduce.[3] That's just the way it goes. For males, however, there is nothing gained by being mounted (that we know of) and they almost always attempt to resist. Females often resist also, so a male cannot use resistance as an indication that he's mounting a male. As mentioned already, female resistance to mating is not uncommon across all kinds of animals, with the most common explanation being quality control. In many species, the way to guarantee a strong and virile father for your offspring is to make it difficult to mate. Females have a reproductive interest in mating with these tough-guy males because their own sons are then likely to take after them and be successful. This sets up a battle of the sexes, so to speak, which in biology terms is called *reproductive conflict*. By resisting, females ensure that they mate only with tip-top males, meaning those that can catch and subdue them.*

How traumatic insemination evolved requires some explanation. The species that use it are all pretty distantly related, which means that it evolved independently and must therefore offer some unique advantages. One hypothesis is that traumatic insemination first emerged as a means to circumvent a mating plug. In many insects—and indeed, other kinds of animals including

---

* In case you are wondering, yes, it has been suggested that BDSM-flavored sex in humans may have its root in this same evolutionary logic. The spectacular commercial success of *Fifty Shades of Grey* speaks for itself. As a close female friend of mine once joked, "Who doesn't love a light spanking?"

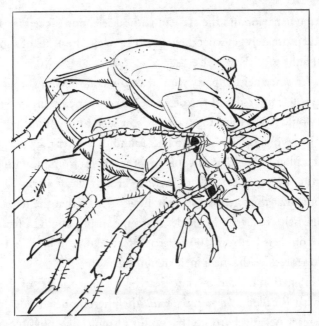

Two bed bugs copulating through traumatic insemination

some mammals—a male leaves behind a gooey plug after he deposits sperm in the female genital tract in order to prevent the possibility of another male subsequently mating and usurping his paternity. Intense male-male competition can kick off an intraspecies evolutionary arms race that, in the case of the bed bug, may have led to traumatic insemination as a way to literally get around the mating plug. It may even have been an accident at first, as these insects aren't known for their precise motor coordination.

Bed bugs, like all arthropods, have an open circulatory system. Instead of blood, they have a substance called hemolymph, which bathes all the cells and organs of the animal without being restricted to vessels. In vertebrates like us, simply injecting sperm into the abdominal cavity of a female would never be successful because we have a closed circulation and our blood is restricted to vessels. Even a tiny cell like a sperm cannot simply swim in and out of the circulatory system. But in the case of hemolymph and an open circulatory system, this crude technique can work. It's possible that, either out of frustration or by accident, male bed bugs began injecting sperm by piercing the abdomen of females. As this strategy found success, the resulting offspring inherited the injecting behavior and the cycle of natural selection kicked in. The poor females still ultimately get what they are after as well: the genes of a strong and successful father.

However, since neither party wins in a male-male bed bug mounting and one party is left injured and the other depleted of sperm, one would think that they would have evolved a way to prevent such futile mountings. Indeed, they may be trying. In 2009, researchers found that, after they eat, male bed bugs emit a pheromone that discourages other males from mounting them.[4] This pheromone was previously thought to be a predator alarm signal. That may be, but what the scientists suggest, however, was that the release of this pheromone helps to alert other males that a mounting would be fruitless and they move on. It is far from 100 percent effective but, in laboratory experiments, it does reduce the male-male mounting.

The explanation offered by the researchers is that the use of this pheromone is what is called a *borrowed signal*. The pheromone already exists and bed bugs react negatively to it. That's a lot of molecular physiology that is already in place, so co-opting this signal for a new purpose requires minimal tweaking. From meaning "Hey, whoever can hear this, Scram! There is predator nearby!" to meaning "Hey, whoever can hear this, Scram! I'm not a female!" is not a big shift, so it's easy to see how this could have evolved, especially since the desired response is exactly the same. This would require only the tiniest tweaking, an evolutionary "piece of cake." For this reason, borrowed signals are common among animals, and even humans use them extensively.*

The fact that it is only partially effective may indicate that this is a fairly recent adaptive innovation and the species is still evolving into it. For the evolution to continue, there has to be a clear selective advantage for both those that release the signal and those that properly respond to it. Those conditions appear to have been met, so it's not obvious why this harmful behavior hasn't been more effectively managed. Evolution of something so simple with such a clear advantage should be quick and easy—in evolutionary timescales, that is.

---

* One example of a borrowed signal in humans: in many cultures, the same gestures that accompany greetings, such as handshakes, bows, or hugs, are often used during reconciliation. This body language signals either potential affiliation or reaffiliation following conflict.

Therefore, I offer another possible explanation. Perhaps bed bugs are well aware of who is male and who is female, and the traumatic insemination that males thrust upon each other is an intentional injury. The males that mount and wound the other males most definitely gain an advantage in doing so, because they are usually in competition with each other. Bed bugs are not particularly social and don't build cooperative partnerships with each other (that we know of). When a male sees another male, particularly a well fed one, all he sees is a potential threat to his future paternity. It should not surprise us that he immediately seeks to reduce or eliminate that threat with the only weapon that he has: his penis.

Furthermore, the supposed pheromone signal that the mounted male gives off may be just what we thought it was all along: an alarm signal, not an attempt to communicate that he is a male. It is possible that the threatened male gives off this signal indiscriminately whenever he feels threatened, whether it's by a predator or another male, like a squid squirting ink or a skunk expressing its anal glands. Or maybe he's attempting to deceive the other male by crying wolf, a so-called *dishonest signal*. Either way, this seems quite a bit more plausible to me because if there was an evolutionary pressure to co-opt an alarm pheromone as a sex advertising signal, why wouldn't that same pressure have already simply led to males evolving the ability to detect who is male and who is female, something that is so universally advantageous that basically all animals can do it.

I tell this story here not because of any obvious parallel between human and bed bug sex, but because it teaches us that the sex lives of even the simplest animals can be far more complex than they initially appear. The use of traumatic insemination for copulation is found in only a handful of invertebrates and incurs costs and injuries for some members of the species that a simpler method of procreation would not. There are attempts to communicate, to resist, and to deceive. There are injuries, strategies, and counter-adaptations, and the scientists that study these phenomena are still working out many details about how these tiny creatures engage in their reproductive behaviors. Even in the simple bed bug, sex is a complicated affair and there is almost always more than meets the eye.

Now consider: What if a male bed bug were strictly heterosexual? And remember that we define *heterosexual* not as being drawn toward sex with those of the opposite sex—lots of orientations have that drive—but as *aversion* to sexual activity with those of the same sex. What advantage would this bring? None that I can think of. But I can think of one big disadvantage: the missed opportunity to get a leg up on other males (literally and figuratively). In a sense, being strictly heterosexual would be like refusing to compete. Therefore, in the big reproductive competition among bed bug males, those that mount other males have an advantage over those that don't. While being open to same-sex behaviors brings advantages, being strictly heterosexual brings costs, at least in bed bugs.

Queer as Duck Soup

As someone who spends the majority of his time in the concrete jungle of New York City, one of the sounds of nature that I find most soothing and hypnotic is the chirping of crickets. In most species of crickets, the chirping is done only by males and is used as a mating call designed to serenade females. If the population is large enough, the sound can be overwhelming, especially so if you picture them as a large crowd of obnoxious dude-bros on the hunt for ladies. Despite the persistent myth that crickets make their mating sound by rubbing their legs together, it is actually generated by their wings, of which they have two pairs. Their forewings have leathery serrations that look like the teeth of a comb. When these are scraped by the hindwings, the resulting vibrations are amplified and resonate through air sacs in the cricket's body. In most species of cricket, females lack the serrations altogether, rendering them silent.*

Like everything else in biology, cricket chirping gets a lot more nuanced the closer we look. It turns out that not *all* male crickets chirp. Some pur-

---

* As an interesting aside, the rate of common field cricket chirping is so tightly linked with temperature that you can accurately use their chirps like a thermometer. If you count the number of chirps in fifteen seconds and add forty, you will get the temperature in Fahrenheit. For Celsius, count the chirps in eight seconds and add five. This is referred to as *Dolbear's Law*, even though Amos Dolbear was actually the second person to stumble upon this mathematical relationship. It should rightly be called Brooks's Law, after Margarette W. Brooks, who first published this phenomenon in *Popular Science Monthly* in 1881, a full sixteen years before Dolbear. So strange that a man gets credit for something a woman did first, right?

sue a different strategy, not unlike the cooperator sunfish males I discussed in chapter 2. These males, presumably, refrain from chirping in order to avoid sending a threatening signal to other males. Male crickets do not just chirp to attract females but also to intimidate other males and warn them to stay clear. Crickets are territorial and will normally fight off an encroaching male, but not if the intruder keeps silent. By refraining from chirping, the cooperator male gains the trust of the territorial male and, it appears, they work together to impress females.[5]

When it comes time for copulation in crickets, the females do the mounting, depositing their eggs onto the genital area of the males using specialized structures on their abdomens. Since females do the choosing, we can only assume that they see value in selecting male-male pairs to receive their eggs. Scientists have long known that crickets with the loudest chirps tend to be the most successful at attracting females, but we now know that females also tend to go for territorial males with a silent helper. These are not conflicting observations, because the loudest chirpers are also most likely to attract a silent male helper. In fact, it may very well be that the presence of a paired partner is more attractive to the females than the loud songs themselves. Maybe all this time male crickets have been singing for other males! That's probably overstating things, but there seems to be a connection among strong singers, helper males, and female choice. There are even conditions in which the silent crickets sire more offspring, on average, than the chirping ones.

At this point, you may be tempted to see a silent male not as gay, but simply as a helper—a wingman, if you'll pardon the pun. This is because of our species-chauvinistic tendency to look for exact equivalents of human behavior in animals, rather than trying to understand animals on their own terms. In order to view an animal as gay,* we expect to see anal sex or something like it, but keep in mind that male crickets don't have external genitalia. Crickets don't have sex the human way *at all*. What we do see in these crickets is same-sex courtship, pair-bonding, and cooperation toward sexual-reproductive goals, and I would say that's pretty gay. More important for the theme of this

---

* Throughout this chapter, my use of the terms *gay, straight, top, bottom,* etc. is purely tongue-in-cheek. I certainly do not mean to conflate the sexual behaviors of crickets, sheep, bonobos, or any other animals with the experience of being a queer human. I just think it's boring to read technical terms like *same-sex oriented* over and over.

book, the silent male-oriented male crickets represent a variant that offers advantages in certain contexts.

Interestingly, there is a genetic component to the nonchirping behavior in crickets, and the wings of nonchirpers are similar to those of females. Because they also father offspring, there is a competition in the gene pool among the dominant-territorial behavior and the silent-cooperative behavior. So, while the noisy and silent crickets get along well in a cooperative fashion, their respective genes are actually duking it out in the grand game of survival. This probably sounds familiar by now. The genius of sexual reproduction is the constant generation of diversity as a reservoir of potential adaptations in an uncertain environment. We think of crickets as noisy creatures by nature, that they couldn't possibly be any other way; but if conditions ever change such that the costs of chirping outweigh the benefits, adapting to a silent lifestyle would be simple. The diversity is already there, waiting in the wings, for natural selection to act on it.

And it has. A species of cricket in Hawaii has recently gone silent, not because they are holding back, but because they have evolved away from chirping due to a new selective pressure: an invasive fly whose maggot larvae burrow into the crickets and parasitically feed on them, eventually killing them when they emerge as fully grown adult flies.[6] Since the flies find the crickets by their chirping, staying silent is a means to escape detection. Given how aggressively and successfully these deadly flies hunt, this is a powerful selective pressure, and scientists have observed that the crickets were able to adapt to a silent lifestyle in just twenty generations.

Twenty generations is a blink of the eye in evolutionary timescales, and way too fast for this adaptation to have emerged *de novo*. Instead, the silent variant already existed in the cricket gene pool within those cooperator males. Of course, crickets and other animals can't possibly *know* what variations might come in handy in the future. Diversity is random, and having the right kind for a new challenge is mostly a matter of luck. In this particular cricket species, the silent version is the result of a genetic variation that flattened the wings and removed the chirping combs. These crickets are probably even better at flying, though this has only been modeled, not directly tested.

Intriguingly, the exact same evolutionary process has occurred twice, on two different Hawaiian islands, separated by vast open ocean.[7] However, totally different mutations underpin the anatomical and behavioral changes on these two islands, making this an example of *convergent evolution*. The same selective pressure (the parasitic flies) led to the same adaptation (flat, nonchirping wings) but by a different genetic route. There are many roads to the same destination.

Okay, that's enough from the invertebrates. For our first vertebrate example, let's revisit the whiptail lizards of the American Southwest that I mentioned in chapter 1. While there are several dozen species of whiptails, a few of them are true biological curiosities (*Aspidoscelis uniparens* and *Aspidoscelis neomexicanus* are particularly well studied). You see, these whiptail lizards don't just occasionally display same-sex sexual behavior; it's the only kind of sex that they *ever* have. These species consist only of females.

How a species of vertebrates can even reproduce with only females, let alone thrive in the harsh desert, sounds utterly mysterious, but the unique evolution of this species has been well documented and we understand in general terms how it happened. First of all, the way that they reproduce is through a process called *parthenogenesis*, in which an unfertilized egg develops into a fully realized and distinct individual. Parthenogenesis is not unheard of in plants and some invertebrate animals, including some roundworms, tardigrades, scorpions, mites, bees, and wasps. Although it is rare in vertebrates, the list of unisexual species is growing and includes representatives of fish, amphibians, and reptiles.

In order to get eggs grow into full-fledged organisms without being fertilized by a sperm cell, whiptail eggs simply double their own chromosomes and then proceed to embryonic development. Remember that animal eggs are all genetically distinct from one another, so these eggs do not produce perfect clones of the mother, even though she is the sole parent. That said, in whiptails, each generation of mothers is produced this same way, leading to repeated contractions of genetic diversity. Similar to a hermaphroditic worm reproducing with itself, this is an extreme form of inbreeding and is thus not a good strategy for long-term success. Since the whiptails seem to be very well adapted to their environment, they are doing just fine for now. If and when the environment changes, I suspect the whiptails are in trouble.

More interesting for the purposes of this book is the copulatory behavior of these whiptails. Since their eggs simply develop into full-fledged

Two female whiptail lizards copulating to induce ovulation

individuals all on their own, one might think they've no need for sex and have ditched it altogether. This is most definitely not the case. In fact, the whiptails still pair up for copulation, and the mounting behavior appears to be essential for the laying of eggs, just as it is in regular ol' two-sex lizards. The females engage in courtship behaviors and, once pair-bonded, they will take turns mounting each other in order to induce ovulation. In reptiles, as in most vertebrates, ovulation is driven by surges in progesterone, and these surges are produced in females when they are mounted, a process called *induced ovulation*. The absence of males in these whiptails has not eliminated the need to have sex in order to reproduce! Even though there is no mingling of genetic material, the females still need to mount each other in order to stimulate the reproductive cycle.

Most, if not all, bird species engage in same-sex sexuality and, as we will see in the next section, this can involve long-term pair-bonding and the raising of chicks. Non-pair-bonding birds also engage in same-sex mounting behaviors. Remember the ruffs from chapter 2? Male ruffs gather in large leks to attract females, but they come in two varieties: the more numerous competitive nest-building males, and the satellite males who are sort of deadbeats that do little work other than puff up the numbers, helping to attract females who prefer populous leks. These satellite males sneak paternity by mounting females when the other males are engaged in their display and courtship competitions. However, interestingly, when females are not around, these males promiscuously pursue mountings with the other males. The first reports of this behavior referred to the satellite males as female mimics (eye roll), but a closer look revealed that the satellite males do the mounting just as often as they get mounted by the nest-building males.[8] In other words, they top as often as they bottom, so these are not female mimics. (Female ruffs never mount.)

As most people know, human beings' closest living relative is the chimpanzee, of which there are two sister species: the famous "common chimpanzee" (*Pan troglodytes*) and the much lesser known "pygmy chimpanzee," more properly known as the bonobo (*Pan paniscus*). These two chimp species are equally related to humans, sharing a common ancestor with us that lived between 6 and 7 million years ago, but they are even more closely related to each other, having been divided into two populations by the creation of the Congo River less than 2 million years ago. (Chimps cannot swim well.)

While common chimpanzees are competitive, aggressive, bullying, and—you guessed it—male-dominated, bonobos are much more cooperative, peaceful, and egalitarian.[9] While they are often called matriarchal, the dominance hierarchy is pretty flat. The highest-ranking bonobos are older females who lead the group through a collaboration with other high-ranking females and males. The top females use their position more to resolve conflicts and maintain harmony than to subdue or coerce others for their own gain. Males earn a high rank because of who their mother is, while females vie for a high station through alliances of mutual aid and support. Yes, higher rank translates into more offspring and certain privileges, but competition is generally through influence and alliances, rather than aggression. We could learn so much from bonobo society.

Bonobos are an incredibly sexual and affectionate species. They use sexual activity as a means of social bonding, conflict resolution, and even just as a greeting. Where other species, including the common chimpanzee, use intimidation and physical aggression to establish rank and enforce social rules, the bonobos use affection and sex.[10] It is also striking how "human" their sex acts are. For example, they are the only non-human animal to engage

Two female bonobos engaged in genito-genital rubbing

in prolonged, affectionate tongue-kissing. They also often have sex face-to-face (missionary style); have group sex; and engage extensively in oral sex and manual stimulation. Although the average bonobo has sex ten or twenty times more often than the average common chimp, their birth rate is no higher, despite bonobo females having a longer fertile period. Why not? Because the vast majority of their sex acts are not procreative in nature. In fact, most bonobo sex is homosexual.

The most common sex act for bonobo females is called *genito-genital rubbing*—or GG, for short—in which two females sit face-to-face and rub their engorged clitorises together. Despite bonobos being about half our size, their clitoris is three times larger and more externalized. (Jokes about men not being able to find the clitoris wouldn't work in bonobo society.) Females use homosexual sex not only for bonding but also as a means to "climb the social ladder." The highest-ranking females are the most sought-after sexual partners, and alliances are established and strengthened through sex.[11]

Male bonobos are not quite as sexual as the females, but they are still far more sexually active than other apes. Groups of males cooperate well together, bonding through sexual activity, especially face-to-face mutual penis rubbing, which they can even manage while hanging in the trees! Because food is plentiful south of the Congo River, there is little food competition. And because pretty much everyone has sex with everyone else, there is little competition for mates, either. And the promiscuous nature of the sex means that all males share in paternity, and the identity of biological fathers is always obscure. For this reason, infanticide is completely absent among bonobos, whereas male common chimpanzees and gorillas are well known to murder the offspring of rival males.

It is important to note one big caveat regarding all this same-sex sexual activity in bonobos. While sex definitely facilitates bonding and social cohesion among the group, especially within specific alliances, bonobos do not pair-bond. In humans and many other animals, sexual activity is vital to establishing and strengthening pair-bonds, which are the key unit for raising offspring. Since bonobos do not form pair-bonds, children are raised communally, with little regard for their biological parentage. Therefore, we will return to bonobos in chapter 5, but for now it is sufficient to say that bonobos, our closest living relatives, are a highly sexual species, and that same-sex sexual activity is literally a daily practice for them. The most direct result of all this gay sex is peace and social harmony.

Recently, the idea that same-sex sexuality promotes social harmony got a huge boost.[12] Researchers in Spain collected published data regarding the sexual behaviors of every mammal species that they could find and organized it into something called a *phylogenetic tree*. Phylogenetic trees represent known or believed evolutionary relationships and are used to track the emergence, spread, and evolution of traits. In this way, scientists can infer if a given trait emerged once and spread, emerged multiple times, or even repeatedly appeared and disappeared over evolutionary time. While these kinds of analyses are a form of secondary research, they allow scientists to take a bird's-eye view of a whole group of organisms and draw far-reaching inferences that can then be tested with direct observation and experimentation by other researchers in the field.

The Spanish scientists found that same-sex sexual behaviors are found throughout the mammal class, no surprise. However, they also found that it clustered in certain kinds of animals more than others. For example, of all kinds of mammals, primate species are among the most likely to engage in same-sex sex. Further, among primates, apes are gayer than monkeys or lemurs.

Both of these results place human beings within a group of organisms with a penchant for gay sex.

Second, and most interestingly, the researchers attempted to correlate the prevalence of same-sex sexuality within mammals with other aspects of their behavioral ecology. In other words, they sought other aspects of the social lives of mammals that may correlate with same-sex sexual behaviors. They found that same-sex behaviors are found preferentially in social mammals, rather than solitary ones. This makes sense because, while solitary mammals have sex mainly for procreation, social mammals use sex for a whole variety of purposes, as we'll see in chapter 5. Animals that engage in sex for various social reasons would have no reason to aim their sexual acts only at the opposite sex, so this makes sense.

Third, the scientists found that species in which the social environment is a cutthroat competition—and I mean cutthroat literally, as in species in which adults are known to fight to the death—are less likely to engage in same-sex sexual behaviors; while the more gregarious and cooperative a species is, the more likely that gay sex is part of their social lives. This too makes intuitive sense because when adults of the same sex are sexual with each other, they are less likely to be hostile and murderous. Sex acts create social bonds, resolve conflicts, and pacify the social environment. The phrase "he's a lover, not a fighter" comes to mind. Or, if you prefer technical jargon, the authors concluded that "same-sex sexual behaviour in mammals is a convergent adaptation facilitating the maintenance of social relationships and the diminishing of intrasexual conflicts."[13]

This result means that the differences between common chimpanzees and bonobos may indeed be generalized across mammals, with sexual activity being key toward creating a cooperative and affiliative social environment, rather than a brutally competitive one. But the question remains: What makes an animal same-sex attracted in the first place?

Gay Rams

Some of the most extensive work on same-sex sexual attraction in animals has been done on sheep. This work was led mostly by Oregon scientist Charles

Roselli, whose interest in the mating behavior of sheep was purely agricultural initially. Farmers have known for some time that some bulls or rams (male livestock animals used for breeding) of various species are very difficult to breed because they show almost no interest in it. These males are desirable in all other ways—large size and easy disposition—but for some reason are stubborn when it comes to impregnating females. This can be very frustrating for a farmer and represent a substantial financial loss. Prized bulls and rams are worth thousands for their prime genetic stock. When they can't deliver, their worth drops to zero.

In 1996, Roselli and colleagues published the first article on "gay rams," as the press reported it.[14] The original focus of their study was rams that refused to mount ewes during an attempted breeding. However, after they noticed that some of the ewe-rejectors would happily mount other rams, they decided to ask if these rams might be "male-oriented." To do this, they created a sexual preference test for rams, which involved placing a ram in an enclosure and presenting him with a restrained sheep for mounting. In each test, the ram is presented with a ram, a ewe in heat, or both, and their preferences in repeated trials is recorded.

In the many rams that they tested, they found all possible combinations: rams that prefer ewes and will never mount rams, rams that prefer ewes but will mount rams if that's the only choice, rams that prefer rams but will mount ewes, and rams that will only mount other rams. They also found that some rams would mount neither ewes nor rams. Later studies would attempt to estimate the size of these various groups, and researchers had little trouble identifying six exclusively male-oriented rams for study, along with seven exclusively female-oriented rams for comparison.

What Roselli and his colleagues found was fascinating. The heterosexual rams had higher testosterone levels in their blood, but also higher levels of the two most active forms of estrogen.[15, 16] The two results that would spur the most debate in the press and scientific community came in 2004. First, the researchers found a small difference in a region of the brains of the gay sheep called the pre-optic area (POA).[17] Second, within the POA, the male-oriented rams have around half as much activity of an enzyme called *aromatase*, also known as estrogen synthase.[18] This enzyme converts andro-gens, sometimes called "male sex hormones," into estrogens, or "female sex

Restraint for experimental testing of
sexual preferences of rams

hormones." And the reason the researchers looked at this specific enzyme in this very specific region of the brain is because this is known to be sex-specific in sheep. The male-oriented rams were similar to females in terms of the activity of aromatase in the POA.

To be sure, the role of sex hormones in the brain is complicated, and little is known about how these hormones affect things like sexual preference, but this did not prevent the press from running away with these results, and an uproar ensued. According to the headlines, the researchers had found how "gay brains are different" and so on. Progressives saw definitive proof that homosexuality is biological and not a choice, while conservatives imagined potential treatments for "victims" of homosexuality. I'm going to sidestep the controversy and bad journalism and focus on the science. The scientifically interesting result published in this article is that, for the first time, we had a hard biological variable, involving hormones in the brain, that discriminated male-preferring and female-preferring male sheep. Could this be indicative of the mechanistic basis of sexual orientation?

Before we explore this further, we must first address an issue that will come up pretty much every time we discuss research on sexual orientation. In the twentieth century, even well-meaning scientists were plagued with prejudices we would now consider simplistic or even homophobic. For example, in both humans and animals, males that displayed male-oriented sexual behaviors were believed to suffer from an abundance of femininity. Their brains, even their bodies, were considered not quite fully male. Proper males mount females, and deviations from that indicated incomplete formation of their masculine bodies and brains. Until very recently, that kind of thinking was baked into the experiments that scientists performed to probe sexual behaviors. And it was even more so in the way they discussed their work.

As far back as the 1950s, scientists proposed that mating preferences in animals, including humans, were determined by the action of sex hormones in the brain. This was not a far-fetched possibility, given that sexual behaviors can be stoked, or inhibited, through the administration of sex hormones. Although early experiments in a variety of species "confirmed" this suspicion, a closer look reveals some pretty lazy thinking and shoddy science underneath all of this work.

To determine if hormones affected not just libido but sexual preferences, scientists pumped animals with hormones. By injecting large amounts of estrogen into young animals, either prior to or after birth, scientists could prevent males from developing the "proper" male sexual behaviors. Similarly, by injecting large amounts of androgens (testosterone and its relatives), scientists could turn female animals into male-mimics, complete with aggression and mounting behaviors. The conclusion they drew from these experiments with sheep and other animals was that "gay" male animals had either too much estrogen or not enough testosterone, and the reverse for gay females.

This was a gross overinterpretation. In the females, the excess testosterone didn't just "masculinize" their sexual behaviors; it also masculinized their bodies. This was not a subtle tweaking of brain chemistry. These females were developmentally disabled, antisocial, aggressive, infertile, and suffered from apparent metabolic syndromes. Their lives were short and miserable. In the males, the excess estrogen also led to developmental problems, both behavioral and medical. The only thing we really learned from these kinds of experiments is that dumping huge doses of sex hormones into developing animals causes a variety of disruptions, including in their sexual behaviors. But these experiments do not logically lead us to a role for estrogen or testosterone in same-sex sexual orientation, because the doses far exceeded what is naturally experienced in a developing animal. This is like concluding that orange juice is toxic because if you force a newborn to drink a gallon of it all at once, she will get violently ill. Orange juice isn't toxic at all provided it is consumed in reasonable amounts by a body that is prepared to handle it. The same is true for sex hormones.

Back to the rams: the discovery and definition of persistently "male-oriented" rams was interesting, because the rams had not been treated with hormones or subject to any manipulation whatsoever. The differences that were measured were the result of their natural development, as far as anyone could tell. Roselli and his colleagues showed that eight to ten of the rams were consistently gay, a figure that is strikingly similar to the portion of the human population that identifies as queer in some way.[19] They also found that the POA of gay rams was overall smaller than that of straight rams, and, once again, similar in size to the POA of females. They also confirmed the reduced activity of aromatase in the POA of gay rams.

Right away, we can see that this result does not conform to the notion that gay rams are just not quite masculine enough. While yes, the smaller POA of gay rams makes them similar to ewes, the aromatase enzyme creates estrogens from androgens. This means that the straight rams would have *more* estrogen and *less* testosterone than the gay rams in their POA. This is the opposite of what was predicted by early experiments with hormone injections. When subsequent work confirmed these findings, demolished was the notion that homosexuality in rams was a simple matter of having too much female hormones or not enough male ones.

The discovery of a size difference in the POA of gay and straight rams might just have been a curiosity, but this sexual orientation-based difference has since been found in rats, rhesus monkeys, and—the strangest animal of all—human beings. In fact, part of the reason Roselli and his colleagues decided to look at the rams' POA was because the British neuroscientist Simon LeVay had previously reported that a specific region of the POA is also smaller in straight women and homosexual men than it is heterosexual men.[20] This kind of research is much more difficult to do in humans because measurement of the POA can only occur postmortem.

There were many other limitations and criticisms of LeVay's work, but when the same finding was reported in sheep, rats, and monkeys, many of the critics fell silent. The POA was suspected by LeVay, Roselli, and others because it has long been known to be involved in sexual behaviors, and in the 1980s it was discovered to be smaller in women than in men, though the size ranges in both sexes overlap considerably.[21]

While work on this area continues to this day and there have been contradictory findings, all in all, it appears fairly solid that, on average, the POA of gay men is smaller than that of straight men and similar in size to that of straight women. What this really means is anything but clear, but since this difference emerges and is fixed long before sexual maturity, most scientists agree that it is evidence for an ingrained, biological basis for sexual orientation. We will return to the topic of possible mechanisms of sexual orientation in chapter 8.

Chuff Chums

Same-sex sexual activity is one thing, but pair-bonding is quite another. This is true for both humans and other animals. Many of the animals who engage in same-sex sexual activity described above either don't pair-bond or they do so very fleetingly. This is because pair-bonding in animals only appears in those that invest extensive parental care in their young. Almost no invertebrates spend effort in parental care, nor do the vast majority of fish, amphibians, or reptiles (outside of protecting unhatched eggs). When it comes to pair-bonding, we're mostly talking about birds and mammals, whose helpless young benefit from, indeed require, the watchful eyes of adult caregivers.

Yes, it appears that the pair-bond exists in the animal world mainly as a social unit for raising offspring. It is a behavior that is driven primarily by attachment hormones, and this hormonal system first evolved as a glue to hold two parents together while their offspring need them. If we assume that both members of the pair-bond are the biological parents of the offspring—an assumption that is not always correct—then both have a strong genetic interest in working together to help them succeed. Good parents lead to successful children, who inherit the genetic basis for good parenting, leading them to raise more successful offspring, and so on. Simple Darwinian logic fits the heteronormative worldview that the primary purpose of marriage is to create and nurture children.

As usual, nature puts many creative spins on this simple logic. One of those is the same-sex pair-bond.

Birds have long been recognized as notorious pair-bonders. Millions of birders watch gleefully as pairs of backyard birds tend to their young each spring. Bird lore is replete with tales of bird couples migrating thousands of miles together year after year. While the reputation is an exaggeration of the reality, many bird species do indeed form stable pair-bonds, and in some species, those bonds can last a lifetime. What we didn't know until recently, however, is that many pair-bonded bird couples are same-sex.

Rarely do scientific articles on animal behavior make international headlines, but late in the first decade of our new millennium, articles about "lesbian albatrosses" could be found in outlets from the *New York Times* to the *Daily Mail*. This was coverage of an article reporting that 31 percent of the albatross couples in a specific colony in Hawaii were female-female pairs.[22] The scientists also noted that the female pairs were just as fecund, or nearly so, as their heterosexual counterparts, and the partnerships were just as enduring year after year. One female couple had been together for nineteen years at the conclusion of the study. Ever since, the albatross has been a symbol of gay coupling in the Western world. In fact, there is a gay bar in my home borough of Queens, New York, called Albatross.

Laysan albatrosses produce very large eggs, and a female can lay only one each year. In the study, the female-female dyads showed an impressive degree of coordination. In most years, each pair laid one egg, with maternity alternating annually, such that, over the years, all of the offspring that they nurtured were the progeny of one female or the other. Presumably, these females engaged in extra-pair copulations with unpaired males in order to fertilize their eggs, and then they cooperatively raised the chick together. The coordination was not perfect, however. Nearly half the time, both members of a female-female pair would lay an egg. In all these instances, only one was incu-

Lesbian Laysan albatrosses in affectionate contact

bated and successfully hatched. The ratio of two parents to one egg seems to be all that Laysan albatrosses can manage.

The authors of the study—a trio of ornithologists from the University of Hawaii—presented the phenomenon as the result of an uneven sex ratio.[23] Prior to bonding, young unpaired albatrosses assemble in sex-segregated groups from the time they are fledglings until age seven or eight. Due to uneven migratory patterns among these groups, the colony on the island of Oahu ended up with more females than males, and the authors claimed that this "shortage of males" was the likely reason for the female-female partnerships. As I will explain shortly, this was in line with common thinking at the time that same-sex pairing in birds is purely a side-effect of what biologists call a *biased sex ratio*.

Color me skeptical. In this colony, females only outnumbered males 59 percent to 41 percent. If we assume that all females prefer mating with males, this would leave only a slight surplus of uncoupled adult females. Even if *all* of the uncoupled females then paired with each other instead of remaining unpaired another year, the numbers would predict that only around 18 percent of the couplings, at most, would be female-female. Instead, nearly a third of all pairs were female-female dyads, and, as always, there were still plenty of unpaired males in the colony. By the math alone, this reasoning is unconvincing.

The authors probably felt the need to present a "reason" for the same-sex couplings because the article, published in 2008, was the first report of such behavior in albatrosses. The lesbian albatrosses on Oahu were seen as an aberration in need of an explanation. But, in my view, the real "answer" to the conundrum is easily spotted in the article itself: female-female pairs were also found in an albatross colony on the nearby island of Kuaui, which didn't have much of a biased sex ratio. It turns out that, while the numbers can vary a lot, female-female pairs aren't unusual in Laysan albatrosses *at all*. They are found in colonies throughout the Hawaiian archipelago. The reason it had never been reported before 2008 is simply that no one checked. Because albatrosses are not *dimorphic*—males and females are indistinguishable—everyone simply assumed that the paired birds were heterosexual.

A biased sex ratio does not fully explain the coupling of female albatrosses, but it may contribute. Scientists have found that same-sex couplings in birds is more common when there is a biased sex ratio. Professor Julie Elie led a research team at the University of California at Berkeley that actually tested this explanation in zebra finches (*Taeniopygia guttata*), a bird known for forming lifelong pair-bonds. By capturing and moving birds to different locations, they were able to artificially generate flocks with skewed sex ratios. When they did, they found that the percentage of same-sex dyads indeed followed suit, although not proportionally so.

But, importantly, the same-sex dyads that the finches formed were no less successful than were the opposite-sex ones, and were no less permanent, even after a more balanced sex ratio was restored.[24] For finches, the important thing is to find a high-quality partner with whom to face all of life's challenges. The sex of that partner doesn't seem to matter all that much. And, like many other birds, the strength of the pair-bond is truly impressive.

In 1998, zookeepers at the Central Park Zoo in New York City noticed that two male year-old chinstrap penguins were performing mating rituals with each other, including the characteristic mating calls and "necking,"

Roy, Silo, and Tango at the Central Park Zoo

nuzzling and intertwining their necks together. The pair, named Roy and Silo, pair-bonded and spent most of their time together. They were even spotted sitting on a rock as though it were an egg. The zookeepers tested them with a fake egg, which they dutifully incubated and cared for, though they became distressed when it failed to hatch.

Eventually, Roy and Silo were given a real egg that another penguin couple had abandoned, and they took turns incubating it until at last their young daughter emerged, later named Tango. Roy and Silo fed and cared for Tango for several months and she thrived. This story inspired many articles, and the story became a model for gay families around the world. More than one person gifted my husband and me with the children's book *And Tango Makes Three*, not long after we became foster parents.[25]

The story of Roy and Silo "broke" in the *New York Times* in 2004 and immediately went viral, much to the surprise of penguin zookeepers around the world. It wasn't the story that was surprising to them, but rather the fact that people found it newsworthy. Same-sex pair-bonding is not uncommon among zoo penguins. In fact, two years earlier, an article in the *Advocate* detailed the "romance" between two male penguins just a few miles away at the New York Aquarium in Coney Island. As far back as 1998, same-sex pairings have been reported among wild penguins in Antarctica.[26] A 2010 study by the Centre for Functional and Evolutionary Ecology in France found same-sex pair-bonding to be "widespread" throughout the various species of penguins and that it occurred in both sexes.[27] In any given colony, it is not uncommon for a quarter of the males to be paired with another male.

You would think that such evidence would dismantle heteronormative prejudices, at least among those who study penguins, but biases are stubborn things. Following the 2010 study that reported widespread same-sex pairings in penguins, the lead author said that these mostly involve high-testosterone males who are "lonely" when there are not enough females around. He went on to say that the same-sex pairings usually only last a few years, and after their dissolution, the next partnership that the penguin pursues is often heterosexual. Indeed, Roy and Silo "broke up" not long after their story went viral, after more than six years together. Roy remained a bachelor until his death, while Silo paired with a female named Scrappy.

Is the breakup of Roy and Silo evidence that the same-sex pairing was unusual? Is the fact that these same-sex partnerships often end in dissolution also evidence of their oddity? Since these formerly "gay" penguins later pair with females, does that mean that they aren't really gay and it was just a fluke or a phase? *Not. At. All.* Here's why.

When the scientist mentioned above described the same-sex pairings as temporary, what he failed to mention is that opposite-sex pairings often end in dissolution as well. Like many birds, penguins pair-bond for a season and then seem to reassess the bond annually thereafter.[28] They usually re-up and remain bonded for the next year, but not always. This is called *seasonal monogamy*, and the annual divorce rate is as high as 50 percent in some species. Each year brings a chance that the couple will go their separate ways and pursue a new pair-bond. In fact, a penguin remaining in the same pair-bond for its entire natural lifetime is rare. So eager were the scientists to paint the same-sex pairing as intrinsically different, they were holding "gay penguins" to a higher standard.

And what about the tendency for penguins to pursue opposite-sex pairings after a same-sex one? For those who hold heterosexuality as the normal preferred state, including some scientists, this speaks to the ephemeral, perhaps aberrant nature of the same-sex pairing. But, yet again, that's a double-standard. We could easily look at this the other way around and say that many penguins in opposite-sex pairings were formerly in same-sex ones, or they will be in the future. Why is one arrangement aberrant and the other natural? Maybe penguins aren't terribly bothered by such distinctions and simply pursue the pair-bonds that feel right each year.

Birds can get even more creative with their pair-bonding. Eurasian Oystercatchers (*Haematopus ostralegus*), a wading bird that lives along the coasts of oceans and seas throughout Europe, Asia, and Africa, are pretty stable pair-bonders. And, like the albatrosses, they are known to occasionally form female-female pair-bonds. However, most often, this is not a duo but a trio with a male. Oystercatcher females are usually territorial and aggressive toward one another, competing over males, who are also territorial and aggressive with each other. When a male and female form a bond and begin to cooperate, their territories will overlap, but not merge, and females will continue to aggressively chase out nearby females.

Except when they don't. Sometimes two neighboring females will simply declare a truce and agree to share a male, whose territory overlaps them both.[29] They don't bond or cooperate, but they tolerate each other, save for the occasional sniping and aggressive display. Slightly less often, however, the two females *will* fully cooperate, form a pair-bond with each other, and merge their nests. And while these are a tiny minority of the oystercatcher groups (only around 2 percent), the paired females preen and mount each other and defend their common territory together with the male that they share. The bonds appear to be nearly as stable as the typical male-female dyads.[30] With three birds to build and defend the nest, with plenty of eggs laid by both females, these oystercatcher "throuples" are successful, much more so than the females who remain aggressive toward one another when sharing a male. The arrangement isn't common, but it works.

Geese are also well-known to form same-sex dyads. In the common goose (*Anser anser*), pair-bonds typically last a lifetime, more than ten years, and as many as 15 percent of dyads are male-male. In fact, scientists often describe the male-male dyads as more tightly bonded, meaning that they spend more time in affectionate contact, grooming, courtship, and sex with each other than heterosexual pairs do. Interestingly, flocks of geese have a dominance hierarchy, with lots of privileges attached to it. Because couples share a ranking, and because males are larger and stronger than females, the male-male dyads tend to be ranked at or near the top of the flock. This means that, in geese, homosexuality brings a very clear advantage. And because geese are fairly promiscuous with their extra-pair copulations, the gay males still father plenty of goslings.

And it's not only the gay males that are at an advantage. One of the first careful, close-up study of female-female bird pairs was conducted with the lesser snow goose (*Chen c. caerulescens*) in response to a prior observation that, in certain years when the number of males is lower than usual, the average number of eggs per nest is actually higher.[31] This confused ornithologists because, while it has long been known that the number of males is largely irrelevant to population growth, no one would expect that the number of eggs laid would actually *increase* when a population is short of males. After all, while males don't lay eggs and one male could cover a lot of females, the constant need for warm incubation means that birds do still require pairs in order to successfully hatch eggs. Males do have *some* function.

So why would a lack of males mean more eggs? It turns out that the number of eggs *per female* did not increase, but the number of eggs *per nest* did, because many of the nests were tended by *two* females rather than just one.[32] When the population is short of males, more birds form female-female dyads, and these dyads lay more eggs than heterosexual ones because, you know, both partners can lay eggs. In a 1989 essay about this research in *Nature* (the most prestigious scientific journal in the world), one of the world's most influential biologists, Jared Diamond, summarized it like this:

> These observations raise the perennial question of why males exist at all at a sex ratio near 1.0. After a male gull has contributed semen, he appears to play almost no role that a female cannot play equally well. . . . Males are useful for protecting the young. Yet other males themselves are one of the main threats in the first place. Further study of homosexually paired female birds may help clarify what, if anything, males are good for—in an evolutionary sense, of course.[33]

That was near thirty years ago! And female-female pairings had been reported in seagulls as far back as the 1970s![34] The data that same-sex pairings are common, stable, and successful in birds has been staring us in the face for decades, but old prejudices die hard.

On the other side of the globe, Australian black swans (*Cygnus atratus*) are also well known for the prevalence of male-male dyads, which are even more common than they are in geese.[35] Up to a third of male black swans seem to prefer pairing with other males. While most migratory birds are socially monogamous, forming stable and enduring pair-bonds, almost none are *sexually* monogamous, something we will explore in the next chapter. Swans, however, are the exception. They are among the most sexually monogamous bird species on record. But the exception to this exception are the male-male dyads, who must pursue extra-pair copulations in order to reproduce. However, in a testament to their fidelity, they do so not by sneaking off to inseminate clandestinely, but rather, by inviting a female to roost with them for a while. They engage in three-way sex, taking turns mounting each other, and then continue in close contact through the laying, incubation, and hatching of the eggs. Even though they are "sharing" a female, these gay swans are just as reproductively successful as their straight counterparts.[36]

The story of why gay swans are more reproductive than straight swans is really the story of why birds pair-bond in the first place: the requirement that eggs be incubated continuously until they hatch places a very unique requirement on the lifestyle of birds. Invertebrates, fish, amphibians, and reptiles all lay eggs, but none of those eggs need to be incubated (although some may need to be guarded). Mammals are at the other extreme and carry around their developing embryos inside their own bodies, so incubation isn't an issue there either. It's really just the birds that have to worry about keeping their eggs warm continuously for about a month. This is why birds are such prolific pair-bonders. Someone has to sit on the eggs at all times, so if it were every bird for herself, she would starve while she did so. Instead, members of a bird pair take turns incubating while the other forages for food to bring back. It's an excellent model of partnership and mutual support. Further, once the egg is laid, there is no need to bother with gender roles. In nearly all bird species, mothers and fathers contribute to childcare, and thus same-sex dyads operate in pretty much the same way that opposite-sex ones do.

This brings us back to the throuples that some swans form. If two swans are better than one, than three swans must be better than two, at least in certain instances. Each bird species has unique challenges when it comes to nest-building, reproduction, foraging for food, etc., so they each develop their own lifestyle that best meets their unique ecological conditions. Under certain circumstances, swan throuples thrive even better than their dyad counterparts, but there are probably good reasons why throuples are nowhere near as common as pairs throughout bird species. For one thing, if there are two males, their paternity is shared, so the female would have to lay *twice* as many eggs in order to match the paternal output of a heterosexual dyad. Eggs are very demanding to create, so that's a big ask for a bird. On the other hand, if there are two females in the throuple, they may lay more eggs than the three birds together can manage.

For our purpose, the key lesson here is that birds have flexibility in their lifestyle and will experiment with various arrangements in order to find how they best thrive. As always, diversity wins.

Sodom's Razor

Having seen all this same-sex sexual activity in all kinds of different animals, we can now return to the question of whether homo/bisexuality is natural and answer it with a resounding *yes*. There is nothing more natural than two animals engaging in a sexual act unencumbered by social or moral conventions. You can bet that they couldn't care less about our sensibilities regarding their sex lives. The vast majority of social animal species—most birds and mammals—exhibit same-sex sexual activity, and this should put to a good and final rest any notion that homosexuality is unnatural, as even the slightest effort toward researching this question brings a plethora of published studies in the scientific literature.

Nevertheless, the scientific study of homosexuality, both in people and animals, has stubbornly clung to the idea that it is a conundrum. While essentially all biologists now accept that same-sex sexual activity occurs and is a natural part of the social behaviors of animals, most continue to approach it as a mystery in need of an explanation. There is an assumption that, all else being equal, a homosexual animal is at a disadvantage in the overall goal of propagating his genes. Given the intense competition we see in nature, any trait that reduces reproduction, even slightly, should be eliminated as natural selection slowly purges it over a few generations. Being attracted to members of the same sex, rather than the opposite sex, would reduce reproductive output, at least slightly, right? That's all it would take for it to be quickly weeded out.

But it hasn't been weeded out. The number of social animal species in which same-sex sexual activity has been sought and *not* found is vanishingly small. There are really only two conclusions that we can draw from that: either same-sex sexual activity brings reproductive benefits that outweigh its costs, or the costs really aren't very great in the first place. Either way, the default position—that homosexuality is at best a quirk and at worst an aberration—must be wrong. Clearly, it is neither. Sexuality in nature is nowhere near as restricted as the traditional human constructions are, and it's time for the science of sexuality to recognize that. It's time for some new thinking.

In one of life's little moments of extreme coincidence, I was sitting at my desk writing this exact section when I saw notification that a paper on this topic was published in the top-tier research journal *Nature Ecology and Evolution*.[37] The article's authors, led by Julia Monk at Yale, pursued a radical approach. Rather than considering same-sex sexual behavior in animals as a puzzle in search of a solution, they approached it as they would any other animal behavior and without any bias about its "costs." They performed what is called a *phylogenetic analysis*, which is a technique designed to reveal when and in which population a certain trait evolved and was passed down to descendant populations and species. By considering which animals possess a certain trait and which ones don't, and considering the known or presumed evolutionary relationships, we can infer the evolutionary history of the trait. This is how we know things like which dinosaurs likely had feathers, even if no feathers have been discovered with most dinosaur species. By knowing who did and didn't have feathers among a dinosaur's relatives, we can make a very good educated guess.

The radical approach that these scientists engaged was simply dropping the assumption that sexual behaviors were originally oriented toward opposite-sex partners in early animals, and that any deviations from that were evolutionary events that needed explanations. Instead, a different assumption was put in place: that there was no sexual "orientation" in the first place, and that specific preferences and orientations would have evolved later, if at all. Making and testing assumptions is a key aspect of the scientific method and the results can be scored on their likelihood, with the general rule that the explanation that requires the fewest assumptions is the most likely. This logic is based on a principle called *Occam's razor*, the idea that the simplest explanation is usually the correct one. As Professor Monk and her colleagues tested various assumptions regarding sexual orientation in animals, they were on the lookout for the simplest and therefore most likely evolutionary scenario.

If we assume that the earliest animals were heterosexual, then we would have to explain how same-sex sexual activity evolved in each of the myriad groups of animals in which we see it. This does indeed seem unlikely. But if we assume that animals were drawn to engage in sex with *any* members of their species, the only behavior that requires an evolutionary explanation is an *aversion* to having sex with certain members of their species. Since most (but not all) of the animals that engage in same-sex sexual activity also engage in some opposite-sex sexual activity, this requires no explanation at all. Most animals have retained the default state of bi- or pan-sexuality, if you will. The animals that require an explanation are those that *only* pursue sex with members of the opposite sex. Therefore, it is not homosexual animals that require an explanation, but rather strictly heterosexual ones.

If you feel that this goes a little too far, please keep one very important fact in mind: early animals did not have different sexes at all. As discussed in chapter 1, the evolution of different sexes in the animal world proceeded in several phases. First there was *isogamy*, when all individuals were the same "sex," even when it came to the creation of gametes. Then there was *anisogamy without dimorphism*, meaning that there were different "mating types" that made self-incompatible gametes, but those gametes looked more or less identical and were otherwise constructed the same way. Then, even when two very different gametes emerged—sperm and egg—the animals that made them were hermaphroditic. Through all of this early diversity in sexual reproduction, the individuals themselves were not sexually distinct in most or all species. There were not separate male or female forms, and so there was no such thing as heterosexuality. Everyone was attracted to everyone else. Why should heterosexuality be the assumed default when, in a sense, bisexuality long predates it?

With that in mind, the evolution of sexual orientation is exactly the opposite of the paradigm that is often presented. Sexual attraction based on the qualities of the other individual, both physical and behavioral, without regard for sex or gender, is the true default situation, and this is closer to what we see in most birds and mammals today. The anomaly that requires an explanation is being averse to having sex with an entire sex or gender because that is less likely, requires more assumptions, and comes with costs. What is to be gained by such aversions?

If we were to put animals on the Kinsey scale—with *0* as exclusively heterosexual and *6* as exclusively homosexual—we would not get the bimodal scale that we observe in humans, with most people clustered near the two extremes. We'd probably get the exact opposite—a normal distribution with most individuals somewhere in the middle. There just aren't many Kinsey-0s and Kinsey-6s among animals. Truth be told, I'm not convinced that there are among humans, either.

But let's hold that thought for now. First we need to discuss all the other purposes and benefits of sex, besides just procreation. Those are the subject of the next two chapters.

Chapter 4
Monogamish

Prior to the late 1980s, it was believed that when birds form stable breeding pairs, they are sexually faithful to each other, at least most of the time. In other words, it was assumed that most birds were *monogamous*. Once a bird couple has formed, they work together cooperatively to build a nest, incubate the eggs, and care for the hatchlings until they can fly off on their own, which means that the couple spends most of their time together, separating only to provision food or nesting materials and other work essential to their shared goal of producing as many shared offspring as possible.

For this reason, birds have been held up as an ideal for coupling, lifelong commitment to family, and sexual fidelity. The expressions "love birds" and "love nest" come to mind. This view of bird monogamy began to change upon the arrival of DNA-based paternity testing for birds, beginning with indigo buntings (*Passerina cyanea*) in 1987 and the Sunda zebra finch (*Poephila guttata*) in 1988.[1,2] One-by-one, ornithologists began subjecting hatchlings of their favorite bird species to genetic scrutiny. And, like an episode of the *Maury Povich Show*, the DNA results came back shouting, "You are NOT the father!"

What the DNA testing showed was that, more often than not, not all of the hatchlings of a given bird pair were the genetic progeny of the male. The specifics vary widely among different species, but what biologists started discovering everywhere they looked is called *extra-pair paternity*, or *EPP*, which is the natural result of *extra-pair copulation*, or *EPC*. How could we, including the biologists who spent their lives studying birds, have been so wrong about monogamy in birds? Because we believed—or I should say *assumed*—that lifelong or seasonal coupling and joint commitment to child-rearing was

An episode of the Murray Ostrich Show

accompanied by sexual exclusivity. In fact, as Professor Sue Carter wrote in "The Monogamy Paradox," "This assumption has repeatedly been shown to be incorrect."[3]

It seems remarkable that biologists were unaware of this until just thirty-five years ago, but a simple literature search reveals the ignorance: searching Google Scholar for "birds, extra-pair paternity" produces exactly *one* scientific article from the years 1977–1987. The same search from 1988–1998, after the introduction of genetic paternity testing, produces 810 articles! Once again blinded by bias, when scientists observed pair-bonding in birds, we assumed that it was sexually monogamous. Only the conclusive nature of DNA evidence forced us to see things as they really are.

What Even *Is* Monogamy?

It's not that biologists were wrong about pair-bonding in birds. We were wrong about what pair-bonding really means because we confused *social monogamy* with *sexual monogamy*. Sexual monogamy is just what it sounds like:

a bilateral sexual relationship, usually preceded by some form of courtship and mate selection, in which sexual activity is mutually exclusive. This sexual exclusivity is what comes to mind for most people when they hear the term *monogamy*, but as we will see, sexual monogamy is exceedingly rare in nature—almost nonexistent, in fact.

Social monogamy, on the other hand, is quite common. Also called *economic monogamy*, this is the tendency of animals, including humans, to pursue dyads that are enduring, marked by mutual aid and support, and that takes priority over all other social relationships. Simply put, social monogamy is two individuals building a life together, most often for the purpose of raising children. They share and defend a home (or a nest, territory, etc.); they share economic resources; they jointly contribute to the rearing of children, usually their own genetic offspring; and, perhaps most important, they do so to the exclusion of others and for a substantial portion of their life history, if not perpetually.

Because we tend to see life through our own lens, mammals often come to mind in discussions about monogamy, but the real stars of social and economic monogamy are birds.[4] Something like 90 percent of bird species form enduring dyads, whereas only 5 percent of mammal species do and 15 percent of primates. The reason for this is the unique way that bird embryos develop: externally, but with warmth required. Reptiles, amphibians, and fish do not tend to their eggs once they are laid, since they do not require warmth. Therefore, dyads are just not a thing among those kinds of animals. Mammals, at the other extreme, carry their developing embryos inside their body, so while partnerships can be helpful, they are not usually required.

Birds, however, are in what I call the "monogamous middle." This is because, in order for bird eggs to develop properly, they must be incubated constantly, and they also must be protected from egg predators. In other words, someone must be sitting on the eggs pretty much all the time until they hatch. And the work isn't finished then, either. The brood continues to need both warmth and protection as their tiny bodies begin to grow. And also, you know, food.

What is a bird mother to do while tending to her eggs and hatchlings for all those weeks? A friend to help with food, water, and defense would sure be helpful. What friend would selflessly help in this way? Why, the other genetic parent of the eggs, of course. Eggs and offspring represent an intertwining of the evolutionary interests of both parents. In fact, in most bird species, the males and females divide the work more or less equally. While one parent sits on the eggs, the other forages for food for them both, and they usually take turns. It is a true partnership in which both parties contribute to the care and defense of the fledglings. This is why monogamous male-female dyads are so common in birds: it is the partnership that is most free of reproductive conflict. A single parent would struggle to stay alive while incubating eggs, but a partnership of more than two would breed reproductive competition, if you'll pardon the pun. Two is just right. Well, most of the time. Variations do exist. (Don't they always?)

The permanency of bird dyads, however, is another story. The lifestyle of migratory birds is highly seasonal. Nests are built and eggs are laid in the springtime; summer is for raising chicks; and fall is for flying someplace warm for the winter. Therefore, the partnerships are often seasonal as well, with about half of bird species displaying *seasonal monogamy*: once the chicks leave the nest, the dyad dissolves, and the cycle of courtship and mating begins anew the following year.

It is true that *some* dyads in *some* bird species are permanent, but the discovery of these lifelong bird "marriages" in the nineteenth century was overgeneralized. Victorians were all too eager to see in nature what they valued in human society: fidelity and permanency. But we now know that the seasonal monogamy of birds is akin to *serial monogamy* in humans, using today's parlance. And, as we will soon discuss, even in staunchly economically monogamous birds, sexual monogamy does not necessarily follow.

For mammals, on the other hand, economically monogamous dyads are just one way to raise offspring, and it is not a particularly common one.[5] There are examples of almost all kinds of mammal parenting relationships you can imagine, from the strong economic monogamy of prairie voles to the group parenting of lionesses. Orangutans and Tasmanian devils are almost exclusively solitary as adults, while sea lions and vervet monkeys engage in *alloparenting*, or cooperative group parenting. Mammal family structures are diverse and flexible, reflecting the arrangement that best fits the particular social ecology of that time and place. But times and places change, and with them, so does the "optimal" family arrangement. As always, the best bulwark against an uncertain future and a changing environment is diversity and flexibility, the enduring lesson of this book. Let's begin by taking a closer look at monogamy in birds.

Avian Hanky-Panky

The discovery of EPP taught us that birds have much more interesting and complicated social lives than we thought. For one thing, it is now clear that essentially all birds, both male and female, and even the most staunchly socially monogamous ones, engage in EPC. If their reproductive interests are already aligned in the success of their shared offspring, why do they do this?

When animal behaviorists seek to understand what motivates a behavior, they must make a clear distinction between the *proximate* goals—the goal, drive, or motivation that the animal is aware of—and the *ultimate* goals—the evolutionary purpose for that drive. Ultimate motives are more complex, more difficult to determine with certainty, and are usually unconscious to the actor. For example, the proximate reason we eat a cookie is because it gives us pleasure to do so, but the ultimate reason we want to eat the cookie is because evolution has shaped our tastes to seek foods rich in calories or other nutrients that were often limiting in the deep past, such as sugar, protein, salt, or certain micronutrients (vitamins and minerals).*

The proximate reason why birds seek EPC is the same reason we all do: they experience sexual instincts and are driven to act on them. Sexual activity with a new and desirable partner tickles the reward center in their bird brains. Just like us, that's all the bird knows. But the *ultimate* reasons why birds seek

---

* Our craving for sweet food evolved when the few sweet foods our ancestors had access to were ripe fruits (and possibly honey), which were less calorific than the domesticated varieties and provided fiber and essential vitamins. Nowadays, the sweet foods at our fingertips provide no micronutrients and are packed with calories, of which we already have an overabundance. For this reason, the human sweet tooth is considered a maladaptation and leads to obesity and type II diabetes. These diseases are increasingly being viewed as evolutionary mismatch conditions, resulting from the stark differences between the modern world and the one in which our ancestors evolved.

EPC are much more revealing about the goals of sex and reproduction and how these shape our sexual attraction. This is worth exploring in birds because some of what we learn may also apply to us.

We must begin by acknowledging that, even though males and females cooperatively contribute to egg incubation and parental care in most bird species, there can still be some differences in the reproductive interests of males and females. Yes, both parties benefit from the success of their mutual children, but each can also pursue genetic posterity in other ways, and these differences can put them in conflict. For example, while a father is away from the nest procuring food for his mate and/or their offspring, he has a clear interest in trying to also sneak in some EPC with any females he finds that are not already incubating eggs or caring for offspring. It doesn't matter if he's only rarely successful because there is so little risk involved. When the EPC *is* successful, it would earn him some EPP at almost no cost because the female is or will be bonded to another male and those two would then invest in the care of the resulting offspring. It's a sneaky way to leave more offspring, so it is no surprise that male philandry is essentially universal in pair-bonded birds.

For females, it's a different story. She cannot just spread her eggs around the way that males can spread their sperm around.* When she has an EPC, she must continue to carry the embryo inside her until she lays the egg, which she must then incubate, defend, and eventually feed. For this reason, females are more choosy than males when it comes to EPC because they must bear the costs, quite literally, of any offspring that results, whereas a male may not have to. There is no free lunch for females.

Or is there? Most bird species do not lay a single clutch of eggs at once, but rather lay eggs repeatedly over a short period of time. Bluebirds, for example, lay one egg per day for up to a week. Out of a desire to have these eggs synchronized in their development, she does not begin to incubate them until she has laid them all. Thus, during the egg-laying period, she is free to pursue copulation and diversify the paternity of her clutch, and this is precisely what she tries to do. In warmer climates, robins can take things even further and

---

* Actually, in some species, she can. This is called *brood parasitism* and involves sneaking one's eggs into the nests of other birds. This strategy can be so successful that some species have evolved to use *only* this method of stolen childcare.

complete several successive cycles of reproduction in a single season, building a nest, laying eggs, and caring for the chicks until they fly away, before repeating the process a second or even third time.

But what is the advantage of having a clutch with multiple fathers? For a male, his EPC could result in his having many more offspring than if he mates only with his paired partner. But for a female, the limiting factor on how many offspring she has is her own ability to produce eggs. Unlike fish, who can squirt hundreds of thousands of eggs in the hopes that a few will survive, a bird can only lay a handful of eggs per season, so she works hard to ensure that each one has a solid chance at success. Whether she mates with one male or one hundred, a female bird will lay the same number of eggs, so why mate with many males, rather than just repeatedly with the best one she can find?

There are two immediate advantages for a female bird to sleep around. First, diversifying the genetic parentage of her offspring is an advantage in and of itself, for all the reasons we've already discussed. Diversity is the key to long-term success in any species. There is no one "best" male for her to find and choose. Rather, each bird has a variety of traits that are good or bad in certain environments at certain times. Yes, good general health and vigor is a constant, but since it is impossible to know what other traits might be best in an uncertain future, variety is the best investment. This is an ultimate motive, of course. Birds, like humans, are not aware of evolutionary reasoning. They are simply driven by their immediate desires and attractions. Over evolutionary time, the drive to pursue diverse EPC has been rewarded in females through the success of their more diversified offspring. The urge to be promiscuous is the product of its own success.

But there is another advantage that females gain when they pursue EPC, even when they don't result in EPP—that is, sex but no conception. One of the first birds in which EPP and EPC was carefully documented through genetic testing was the dunnock (*Prunella modularis*), also called the hedge sparrow, a very common, if unremarkable, backyard songbird native to Europe and Western Asia. Because this was the late 1980s, researchers were surprised to find that many dunnock nests produce broods with a variety of different genetic fathers. This showed that the females had engaged in EPC. But in later studies, these and other researchers observed something they hadn't noticed before: while their mates were away foraging and the females were incubating the eggs or hatchlings, other males would occasionally arrive at the nest and share food with the females.[6] And who were these generous males? You guessed it. They were the ones that she had previously mated with.

By having sex with males beyond her pair-bonded mate, a female dunnock is able to procure additional support for herself and her offspring. These extra-pair males don't provide as much support as the primary partner does, but they don't have as much of the paternity either, and most of them have their own primary mate to whom goes most of their support. So what's going on with all of these secondary sexual relationships?

Even today, most researchers consider this phenomenon deceptive and discuss it as a moment of reproductive conflict. In other words, they consider this "cheating," and the result of the old "battle of the sexes" that has dominated how biologists think about animal behavior for the past few centuries. If you can believe it, another "scientific" term for EPP is *cuckoldry*,* and scientists use this term in published research to this day. As you might suspect, I find this cynically anthropomorphic, and unnecessarily so. A simpler explanation is available to us.

By mating outside their pair, both male and female "cheats" are actually just creating an evolutionary hedge investment. Pair-bonded birds put most of their genetic eggs in the basket with their primary partner, if you'll once again pardon the pun. Like everything else in life, mate choice is a gamble, and as any gambler or investor will tell you, it's often wise to make a side bet. In the event that your main bet doesn't go as you had hoped, the side bets can keep you in the game for a while. For these birds, the reproductive side bets don't cost them very much but provide an additional opportunity for fitness—that is, reproductive success. Labeling this as "cheating" or "cuckoldry" implies not only that it is deceptive, which it may or may not actually be, but also that it is underhanded and immoral. This type of moralizing has no place in the analysis of animal behavior. And besides, the birds may be perfectly aware of what is going on.

---

* The word *cuckold* comes from the French word for *cuckoo*, a bird that lays its eggs in the nests of other birds. This makes no sense linguistically as it is the female cuckoos that are the brood parasites, while *cuckold* specifically refers to the *husband* of an unfaithful wife.

That said, this does set up a conflict of interest that must be managed. While a paired female loses nothing when her mate has sex with another female, she does stand to lose something if he later shares food with that female. Similarly, a male whose mate has engaged in EPC may very well end up contributing resources toward the success of other males' genetic offspring. So, while both males and females are interested in pursuing EPC, they both have an interest in trying to minimize their partner from doing so. And they take action accordingly.

For example, it has been suggested that part of why a roosting female accepts the advances of another male is to ensure that her primary male returns as quickly as possible, rather than spending lots of time pursing EPC. In jackdaws (*Corvus monedula*), a monogamous species in the highly intelligent corvid family, roosting females pursue EPC that does not usually result in EPP. It is not clear if the females are just not fertile or if they have a mechanism that prevents successful insemination, but what is clear is that when they are left alone, these females permit interloping males to mount them and emit their distinctive, and loud, copulation calls.[7]

The fact that these birds are loud in bed is precisely the point. Upon hearing those calls, the paired male quickly returns to chase out the intruder. He then immediately copulates with his mate to try to replace any competing sperm with his own. The best way to get her man back where he belongs is to threaten him with the possibility of another man taking his place. There's drama, for sure, but no evidence of deception. In fact, she seems to *want* him to know what she's been up to so that he will come home to where his bread is buttered.

The conflict over EPC in jackdaws introduces us to a behavior that is all too common in both birds and mammals: *mate-guarding*. This is when one member of a pair, invariably the male, attempts to physically interfere with any contact between his mate and members of the opposite sex. When paired male jackdaws hear copulation calls coming from the vicinity of their nest, they rush back home to chase out the suitor. Researchers have observed a pretty good correlation between the threat of EPC and the time spent mate-guarding in jackdaws and other bird species.

Jackdaws are another good example of how the reality gets more complex the closer you look. They have long been recognized as one of the most monogamous bird species because their pair-bonds are almost always lifelong.

Genetic testing later revealed that EPP is rare, so researchers assumed that they were also sexually monogamous. However, per the study mentioned above, published in 2020, we now know that jackdaws engage in quite a bit of EPC; it just doesn't lead to EPP for reasons we don't yet understand. The mere threat of EPP seems to be enough to get the male's attention and draws him back to guard his mate. Thus there is a third form of monogamy that we must consider: *genetic monogamy*. Jackdaws are economically monogamous; they are also genetically monogamous, even though they are *not* sexually monogamous. (I told you monogamy was complicated!)

Let's now take a look at monogamy in mammals. Just as with birds, social relationships in mammals reflect the ecology of the species. If monogamous dyads make sense for the raising of offspring, that's what they do. If not, they don't.

Mammal Families

While bird sexual behaviors are interesting and informative, the sheer diversity and variability of reproductive and sexual relationships in mammals surpasses that of any other group of animals, perhaps more than all others combined.[8] Among mammals, there are species that are monogamous, polyandrous, polygynous, solitary, and cooperative (multi-male and/or multi-female groups that are socially bonded together but without bonded dyads).[9] Within each of these, there are also various permutations and exceptions. It almost seems as though every mammal species has its own unique type of sexual and parenting relationships.

There are two key reasons why mammals show such striking diversity and complexity in their social relationships. First, phylogenetic analysis predicts that the earliest mammals were solitary.[10] We can think of this as a sort of "blank slate" when it comes to sociality. With no hardened social behaviors baked in, myriad social structures developed from the ancestral blank slate and then evolved based on the environments and survival pressures in which the various new species found themselves. Among vertebrates, mammals may be uniquely flexible in their approach to social relationships, and this may be why mammals have become so dominant. Sociality brings a lot of advantages, especially when it is flexible and adaptive.

Mating System	Definition	Mammal Examples
Promiscuity	No social bonding among mates	The majority of mammals
Solitary	Little or no social bonding among adults.	Orangutan, Tasmanian devil
Seasonal monogamy	Close pair-bonds in annual cycles	Muskrat, beaver, Arctic fox
Lifelong monogamy	Permanent pair-bonds	Prairie vole
Harem (polygyny)	Several females bonded with one socially dominant male	Lion, red deer, gorilla
Resource-defense polygyny	A single dominant male mates with several females and limits access to food resources accordingly	Uganda kob
Lekking	Groups of males congregate to both attract females and compete with each other, usually through displays	Some walruses, seals, and sea lions, hammerhead bats
Polyandry	A single dominant female mates with several males	Marmosets, saddle-back tamarins

Table One: Animal mating systems.[11]

Although they existed in the shadows long before, mammals experienced an explosion of diversity beginning 65 million years ago. The extinction of the dinosaurs—which dominated the landscape and seascape for nearly 200 million years—opened up new ecological niches all around the globe, and the previously insignificant mammals quickly took their place. As they did so, they radiated into the major groups we know today, including rodents, carnivores, ungulates (hoofed mammals), cetaceans (whales and dolphins), pachyderms (elephants and their relatives), lagomorphs (rabbits and hares) and, of course, the primates. Each of these groups developed their own unique way of life and the social structures to help them navigate that life. Within each group, further diversity emerged.

The second reason that mammals exhibit such interesting diversity in social structures is because of their unique reproductive physiology involving internal gestation. With the exception of monotremes (the platypus and the echidnas), mammal embryos develop inside the body (or pouch) of the mother. This takes differences in maternal and paternal investment to an extreme. Mammal mothers have literally no choice but to invest an enormous amount of resources into their offspring, but the same is not true of fathers. Of course, in some species, fathers are doting, attentive, and invest just as much as mothers do, even more in some cases. But in many other species, mammal fathers are complete deadbeats. Most are somewhere in the middle.

Interestingly, despite the solitary lifestyle of their ancestors, most present-day mammals have an elaborate variety of social relationships, beginning with the most intimate relationship in the entire animal world: nursing. Were we not so accustomed to it, the notion that an animal could synthesize food with her own body would seem like magic or science fiction! But that is exactly what mammal mothers do, and it is a degree of parental care that is unmatched in the animal kingdom.

The majority of mammals are promiscuous,* meaning that they do not form pair-bonds in any way. Importantly, this does not mean that they don't experience social bonding and attachment; it just means that there isn't any special kind of bonding between breeding pairs. For example, brown rats (*Rattus norvegicus*), the most common rat pests in North American cities) live in large tightly knit social groups. There is a dominance hierarchy and extensive prosocial behaviors such as grooming and food-sharing. Laboratory experiments have even revealed that this much-maligned species experiences empathy and engages in impressive displays of altruism. Rats have individual personalities, form stable friendships, and will go to great lengths to help a friend in distress, even at a cost to themselves.

But when it comes to mating, there are no dyads. While most friendships are male-male or female-female, sex is a total free-for-all. Anyone can mate

---

* The word *promiscuous* can also refer to EPC—that is, social monogamy without sexual monogamy. A 2018 study found that the term was used very inconsistently in the scientific literature. It is also loaded with human cultural baggage. For example, another study found that scientists were significantly more likely to label the same behavior "promiscuous" for a female than for a male. Shocking.

with anyone else, although, as is so often the case in mammals, the females can be choosy. Because rats are prolific reproducers and often stay close to the social group into which they were born, inbreeding is a constant threat to their genetic fitness. Therefore, it is not surprising that, for female rats, one of the most attractive features in a male is novelty. The new guy on the block is highly desirable because, as always, genetic diversity is a key goal of reproduction. Therefore, in this context, pair-bonds would be a detriment to an individual's evolutionary fitness, so it makes sense that the rats would have evolved away from them.

At the other extreme are the prairie voles (*Microtus ochrogaster*), who, at least on the surface, don't seem all that different from rats. But boy, are they. They are the most monogamous mammal species that we know of, meaning they have the strongest preference for their bonded partner over other social partners.[12] So strong is their monogamy that they have become the model organism for studying monogamy, and I will attempt to summarize decades of research on this animal in the next section.

Curiously, most of the prairie vole's closest relatives show no pair-bonding whatsoever, let alone strong monogamy. In fact, there are 155 species of voles spread across nineteen genera, but only a handful are monogamous, and they are randomly distributed throughout the vole family. And the two closest relatives of the vole group are muskrats and lemmings. Within a breeding season, muskrats are sexually and socially monogamous. The many species of lemmings, on the other hand, show a wide variety of mating types, from the monogamous steppe lemmings to the very promiscuous Norway lemmings. And brown lemmings are solitary!

The diversity of mating systems in voles and lemmings reflects the larger diversity of pair-bonding arrangements throughout mammals. There is something special about mammals that breeds˙ diversity when it comes to pair-bonding and other social relationships, and I have already alluded to what that "something special" is: pregnancy. The unique evolutionary pressures on mammals have led to the evolution of unique social structures. It's worth exploring some of this diversity as we approach the topic of our own relationship with monogamy.

---

*   Okay, even I agree that my puns are getting out of control.

Part of the reason why prairie voles are often held up as a model of monogamy is because monogamy is pretty rare in mammals. The most recent survey found that 226 mammal species form socially monogamous dyads for at least a breeding season. That's about 4.5 percent of all mammals, meaning that almost 95 percent of mammals are not monogamous in any way—the exact opposite of the situation in birds, in which 90 percent are seasonally or perpetually monogamous. Of that 4.5 percent, about two-thirds are monogamous for more than a breeding season, bringing us to the oft-cited figure of 3 percent for the portion of mammal species that form lifelong pair-bonds. What's most interesting is that the various monogamous species are spread widely throughout distantly related mammals, rather than clustering together in a closely related group.

As mentioned above, muskrats (*Ondatra zibethicus*) form seasonal pair-bonds that are similar to those of migratory birds.[13] In the spring, these aquatic rodents stake out and defend territories and compete for mates, with a great deal of injuries and even death resulting from all the fighting. Once a tightly bonded breeding pair is formed, they cooperate to build an elaborate "house" on the bank of a river, stream, or lake, with the entrance accessible only underwater. As the pups are born and reared, the family shares the home for the season and cooperate to protect and maintain it. The breeding pair is prolific, producing two or three litters each year and up to twenty offspring. As winter arrives, the family dissolves and everyone goes their own way. This is a classic example of seasonal monogamy in mammals.

Monogamy and promiscuity aren't the only options, however. The mighty African lions (*Panthera leo*) employ the polygynous mating system known as a *harem*.[14] In the Tsavo lions (the most abundant lion subspecies in Southern and Eastern Africa), a harem consists of one dominant male accompanied by several female mates, forming a tightly bonded social group called a *pride*. Prides are fascinating but rife with misconceptions. For one thing, although male lions are often the ones we most fear, these powerful animals don't actually hunt very much. Within a pride, the females do almost all of the hunting, and they bring back their catch to share with the male, the juveniles, and their "sister wives." While there is always some rivalry among the females, and competition within a dominance hierarchy, the females cooperate with each other for the most part, and the male mates with all of them more or less equally.

The sister wives are almost always genetically related and are often literal sisters.[15] This makes the offspring cousins as well as half-siblings, because they are all children of the same male. Interestingly, the females cooperatively contribute to the parenting of all offspring. This is called *alloparenting* and is observed in only about 3 percent of mammal species, mostly primates, and it is virtually unheard of outside of mammals.[16] But it makes sense that we would find alloparenting in lion prides because of how related all of the females and offspring are. Lionesses within a pride will protect, feed, and even nurse any and all kittens, regardless of parentage.[17] In some other species, there is some preference for one's own genetic offspring, but the strength of this preference varies and is tied to how related the sister wives are likely to be. In sperm whales, for example, there is not much allonursing, and mothers generally provision food for only their own older children, but the mothers do take turns collectively "babysitting"—that is, protecting all youngsters from danger while some of the mothers dive very deep for hunting.[18]

With the female lions doing all of the hunting *and* all the child-rearing, what is left for the male lions to do? Not a whole lot. They really are the deadbeats of the cat family and spend the vast majority of their time sleeping, sometimes over twenty hours per day, and this begs the question: Why are they so powerfully built? Male lions are around 50 percent larger than females, but their strength is seldom used to take down prey. Instead, males are built big and strong for one primary purpose: fighting each other.

Often, human men look at the harem lifestyle with envy, but a closer look reveals that it's not all it's cracked up to be. Sure, if you're the alpha male of a pride, you get most of your food procured for you and have to do little else but sleep and have sex, but this comes at a tremendous cost. Occasionally, you have to endure a fight to the death with another male, and even the strongest will eventually lose to a challenger. Being violently killed by another male is the most common way that lions meet their end.

In fact, as young lion kittens are weaned, grow, and become adolescents, the males are in a precarious position. They switch from being a protected child of the male to a potential challenger. Eventually, they must leave the pride, either by choice or because they are pushed out by their father. And because this happens before they are strong enough to mount a challenge to another male, they begin a solitary period when they continue to grow and develop in preparation for the day when they will challenge another male (usually not their father). Most will fail. The harem lifestyle means that the majority of males lose their challenge and die as frustrated virgins. It's not a cushy life for most males, and even those that are successful must always be looking over their shoulder.*

Along the same lines, whenever a male is deposed by a younger, stronger interloper, among the first things he does is to kill all of the younglings in the pride. This prevents his new harem wives from spending any of their time and energy on another male's offspring and also brings them into estrus so that he can begin impregnating them all. The lion is majestic, but he is also brutal, and so is the harem lifestyle.

The reason why lions form such tight-knit social bonds among breeding pairs but rats do not is completely in line with their respective social ecologies. In rats, reproductive rates are very high, little parental investment is required, and inbreeding is a big danger. Therefore, tight social bonding to a breeding partner would bring little benefit. In fact, pair-bonds, and the jealous mate-guarding that comes with them, would actually harm a rat's chances of long-term evolutionary success because they would miss out on the chance to diversify the genetics of their offspring. This is a selective pressure *away* from pair-bonding. In lions, however, generation times are much longer, so reproduction is a slow and steady process that requires a great deal of parental investment. Kittens nurse for at least six months and do not reach maturity for three years. Promiscuous breeding would do little more than generate many kittens who are unlikely to succeed. Instead, by forming a tight-knit pride with a high degree of genetic relatedness, a lion and a small group of lionesses can cooperatively invest their energy and give each kitten a strong chance of

---

* As mentioned in chapter 3, in other subspecies of lions, dominant males often accept a second male as a helper. These beta males are submissive to the alpha and assist him in fighting off any challengers. In return, they are rewarded with a small share of paternity. There are even lion populations in which prides are ruled by a small coalition of up to four males. The partnerships and coalitions are especially useful to a newly mature male hoping to take over a pride, but once again, they cannot turn their back. It's not uncommon for bonded males in a coalition to turn on each other. In the grand scheme of evolutionary success, paternity is everything.

surviving to maturity. The sexual relationships of rats and lions make sense for the way each species advances their long-term success.

Lekking is a mating system we first introduced in chapter 2, in the section about ruffs. Leks are not uncommon among birds but very rare in mammals.[19] However, three distantly related groups of mammals engage in lekking: a few species of pinnipeds (walruses, seals, and sea lions), three species of antelope, and one species of bat, the hammerhead bat. Leks are groups of males that congregate in order to collectively attract females but then compete with each other for the attention of females once they arrive. Competition is not usually combative, but rather in the form of various kinds of courtship displays.

California sea lions (*Zalophus californianus*) employ an interesting mixed lek system and congregate in large sex-segregated breeding groups called *rookeries*.[20] Within these, males engage in competitive courtship activities that are a mix of display and territory-defending. The males "fight" one another for territory within the rookery, but with little direct combat. Instead, they employ courtship rituals that are mostly directed at each other, not the females. These rituals include vocalizations, dances, posturing, head butts, lunging, and, yes, mounting. The competitions can last several weeks, completely consuming their time and energy.[21] The sea lions don't even feed during this time, instead relying on their substantial deposits of blubber. Defending a territory is mostly a matter of intimidation, and the feature that earns the most paternity is endurance. Larger sea lions compete better than smaller ones, not because of strength, but because of their ability to last longer while fasting. Being fat is more important than being strong, and those with the most staying power can sire up to four or five pups in a given year.

The female sea lions also congregate together, and before they begin to mount the males, they spend plenty of time mounting each other.[22] Because sexual activity is required for ovulation in some mammals, it is possible that the female-female mountings help bring the females into their fertile phase as they prepare to allow males to join the fun. Or the females may be "practicing" their sexual actions with each other before the breeding begins. Or perhaps the mountings are some kind of dominance competition, alliance building, or other social structuring that takes place through the sexual activity. We don't really know what kind of ultimate purpose the female-female mounting serves, or even if there is one, but the sea lions' point of view is probably as simple as that they enjoy sex.

While a male sea lion will mate with any female he can, the females are more selective and observe the male-male competition for a while before they begin to grant access.[23] Females tend to go for males who command the largest and most desirable territories. Direct access to water, for example, is key to the desirability of a territory because, while the male sea lions don't eat during the mating season, they must drink often. Aquatic animals are not adapted for water deprivation and dehydrate quickly, so the males who can secure the largest territory with good access to water fare the best with the females.

It seems odd that territory matters so much to a female sea lion's mating choice because the territories are useless beyond the mating season. Most likely, territories are used as a proxy for overall health and vigor. A male that can outcompete and outlast other males in a coveted territory, while going without food for weeks, must be large and healthy, and that speaks to good genes overall. So, rather than inspecting the males themselves and attempting to determine who is the strongest and healthiest, females can simply scan their respective territories and judge the quality of the male based on the quality of his territory. It's a pretty good system.

Proxy systems for judging the quality of males are found in many other mating systems as well and is a very common aspect of how sexual selection plays out in mammals, and sometimes in birds too. For example, the enormous tail of the peacock is purely the product of sexual selection and serves as a proxy system for judging male "quality." It does nothing to aid in their survival; in fact it's quite an impediment to lug that enormous thing around. So what's the purpose? To impress the peahens. But *why* are they impressed? From their point of view, the proximate motivator is simply attraction. Peahens are attracted to large and beautiful tails. But the ultimate motive, the evolutionary reasoning, is that the tail is a proxy indicator of health and vigor. In order to survive with that enormous burden attached to his butt, a peacock must be healthy and strong indeed. The biology jargon for this concept is the *handicap principle*.

The same is true for the antlers found in most deer species and their relatives. Some use them for fighting each other, others purely for display, but the point is the same: the antlers aren't for defense or hunting or any other survival-related purpose; they are purely for winning access to females, either through female choice or direct combat. The gigantic antlers are a hefty encumbrance, not to mention costly to produce, and that's exactly how they serve their purpose. Only a strong and robust deer could survive with such an impediment in place most of the year. Since we see these proxy quality indicator systems in so many different kinds of animals, the benefits must outweigh the costs. Indeed, the strong force of natural selection on the genes of these animals keeps any harmful mutations from spreading through the population. Because genetic material gets shuffled during sexual reproduction, the strong selective force placed on males improves the genetic quality of the whole species, males and females. Once again, it's not a bad system.

Mammals, Moms, and Monogamy (Oh My!)

As promised, I shall now return to the adorable prairie vole (*Microtus ochrogaster*). As the model of mammalian monogamy, these are the animals in which scientists have carefully worked out the hormones and neurocircuitry that underpins attachment and social bonding in mammals, including humans. This work has revealed that the uniqueness of mammal reproduction—pregnancy, live childbirth, and nursing—is actually the evolutionary root of all social bonding, including and especially bonding between sex partners. This is because the key hormone for both birthing contractions *and* the delivery of milk is the very same hormone that facilitates social bonding in mammals. That hormone is called *oxytocin*, a powerful neuromodulator that, together with a closely related brain hormone called *vasopressin*, is key to monogamy.

In a female who has recently given birth, oxytocin is released from her brain whenever she feels suckling on her nipples. This then directs the mammary glands to begin ejecting milk. But that is not the only input that can trigger oxytocin release. Hearing her baby cry can also trigger milk ejection in a new mother. In fact, this trigger is so sensitive that nursing women will often begin leaking milk when they hear *any* baby cry. My mother tells a story of when I was a nursing infant, she had to run out of the grocery store because the cries of a nearby baby produced visible wet spots on her blouse. (Sorry to embarrass you in yet another book, Mom.)

In addition to milk ejection, oxytocin is also responsible for the feeling of attachment that one mammal feels for another, and this is surely not a coincidence. Part of the loving, protective attachment that a mammal mother feels for her child is due to the powerful effects of oxytocin on her emotional state. Evolution, in her ingenious way, made use of the same hormone for both nursing and mother-child attachment. And once this oxytocin-attachment system was in place in early mammals, evolution was then able to exploit this system for other kinds of attachments that could benefit mammals, including the attachment that co-parents feel for each other in species that pair-bond. In short, oxytocin is known as the attachment hormone and, as such, it plays a pivotal role in monogamy.

In prairie voles, the preference for bonded partners can easily be measured and manipulated.[24] Oxytocin and vasopressin are released during sexual activity, and together they promote bonding and attachment between the partners. Each time a bonded pair have sex, the resulting release of these hormones strengthens the attachment. The connection between oxytocin/vasopressin and monogamy has also been observed in other pair-bonding mammals, including primates such as tamarins, titi monkeys, and humans.

Some interesting experiments have demonstrated the power of oxytocin. For example, blocking oxytocin or vasopressin causes voles to temporarily "forget" their preference for their bonded partner and become socially and sexually promiscuous. Once this treatment is stopped, however, the hormones return and the voles go back to their spouses. Oppositely, when given an extra boost of oxytocin and vasopressin, voles will bond even stronger to their mates and spurn all contact with unfamiliar members of the opposite sex.[25]

This phenomenon also works in humans! Oxytocin can be delivered with an intranasal spray, which has allowed scientists to study its effect on human pair-bonds. When scientists give pair-bonded men a dose of oxytocin, they tend to keep more distance from unfamiliar women in a social setting than they do otherwise, an effect that is *not* seen among unattached men. The extra oxytocin strengthened the monogamous attachments to the exclusion of potential interlopers.

Pair-bonded prairie voles

This led to some experimentation with oxytocin in the context of marriage counseling, but these efforts were quickly halted when some of oxytocin's other effects on the human mind reared their ugly head. Men given extra doses of oxytocin may indeed feel more attached to their partners, but they are also more prone to in-group identification (racism and xenophobia) and to feeling threatened with the loss of their attachment (jealousy), sometimes to the point of violence against their partners or perceived threats.[26, 27] Powerful though it is, oxytocin is no easy fix for marital problems.

Oxytocin is key to all forms of social attachments in mammals, not just sexual ones. For example, wolf packs are tight-knit social groups held together by oxytocin-mediated attachments.[28] Most often, wolf packs are nuclear families consisting of a mated pair and their offspring. Obviously, the breeding pair have sex and this promotes their bond, but inbreeding is not common in wolves, meaning that wolf parents do not copulate with their children, nor do siblings copulate with each other. Thus, other aspects of their close affiliation—probably affectionate physical contact—must stimulate the release of oxytocin to promote their social bonds.[*29]

Interestingly, wolves have taught us how other hormones are involved in attachment and social bonding.[30] Specifically, researchers have discovered that when a pack member gets separated, the remaining members will howl to try to guide the lost wolf back. However, the ones most closely affiliated with the lost wolf do a lot more howling than the others, and the hormone cortisol appears to drive this behavior.[31] Cortisol is the main long-term stress hormone (as opposed to short-term stress mediated by adrenaline and others) and is released during times of social stress. When we miss the person to whom we are attached, cortisol surges and we do what we can to try to get them back.

As a side note, wolf packs are also a good example of the benefits of social living over going it alone. Since wolves do sometimes hunt as a pack, it was long believed that cooperative hunting was the main benefit of their close social affiliation. Large, cooperating hunting packs can more successfully bring down big game such as bison and moose, so the thinking went. But this turned out to be false. Scientists have consistently observed that breeding pairs and even lone wolves have higher success rates than do hunting packs, even for large game. If cooperative hunting isn't the benefit of the pack behavior, what is? Food sharing. When wolves are successful in hunting or finding food, they share with their pack members, even among nonrelatives if there are any. Since wolves prey on very large animals, pack living means that only one wolf has to be successful for all to eat. In this context, the value of social living stems not from cooperation or division of labor, but from the sharing of resources.

But back to the voles and oxytocin: ever since the discovery that oxytocin is a chief mediator of pair-bonding in these rodents, they have been held up as a model of monogamy and marriage. As recently as 2014, a story about voles ran in the *Smithsonian Magazine* under the title "What Can Rodents Tell Us about Why Humans Love?"[32] This has long vexed the world's expert on

---

* In fact, it has been hypothesized that oxytocin may have played a key role in the domestication of dogs from their wolf ancestors. As human-companion animals, dogs are unique in that they form close social attachments to members of another species (humans), and these attachments are mediated, in part, by oxytocin. It may be that the oxytocin-mediated social bonding of wolf packs made them uniquely poised to evolve as human companions.

monogamy in voles, Dr. Sue Carter, mentioned above as the author of "The Monogamy Paradox." Professor Carter was the lead author on the first studies of monogamy in voles, as well as the later studies that demonstrated the role of oxytocin and vasopressin. All along, her work with voles has made clear distinctions between social and sexual monogamy, but these distinctions are often missed in the press coverage of her work.[33]

As far back as 1990, Carter's study of vole behavior distinguished between sexual preferences and social preferences. In the very first study, the sexual promiscuity of voles was made clear by DNA fingerprinting: female voles frequently gave birth to litters of mixed parentage. It's right there in the data of the published paper! And yet, most people, even most scientists, saw only what they wanted to see.

Worse still, as Carter and others have worked to set the record straight, the underlying reality of prairie vole monogamy simply gets tossed aside. The most up-voted comment in a recent news article about the voles was, "So they're not really monogamous, lol." *Yes, they are!* Prairie voles show strong social monogamy. They form lifelong dyads, usually heterosexual, and they strongly prefer their bonded partner for play, food sharing, grooming, and other affectionate contact. Most important, they co-parent together! Both bonded partners contribute extensively to parental care of "their" children, even when some of them are not the genetic offspring of the male partner. What matters for the monogamy of voles is how much time they spend together as friends, allies, co-parents, and sex partners, not who else they do or don't have sex with.

Couldn't the same be said about human marriage? Of all the various ways that married couples enjoin their lives together, why does the exclusivity of one behavior—sex—seem to matter more than everything else put together? (I digress.)

What is most interesting about the research on monogamy in voles is, once again, the complex influence of neurohormones on behavior. Oxytocin and vasopressin are released during sex, which helps create and strengthen the pair-bond. But if voles are promiscuous when it comes to sex, how are their pair-bonds so specific? Clearly, oxytocin and vasopressin aren't enough, and pair-bonding is complicated. Professor Carter and others have discovered a role for many other neurohormones as well, including dopamine, estrogens, androgens, and cortisol.

In fact, in a very dramatic experiment, scientists were able to partially transform the males of a nonmonogamous species of vole into monogamous males simply by altering the expression of a single gene: the vasopressin receptor.[34] Interestingly, vasopressin does its monogamy-promoting work in the dopamine reward pathway of the brain. In other words, social monogamy is at least partially driven by the same kind of "pleasure" sensation that we get when we satisfy hunger or thirst, have sex, or use cocaine. For species that have the neurocircuitry in place, monogamous pair-bonding feels very good. So, in a sense, the "drive" for pair-bonding works similarly to the drive for food and sex. In the proximal sense, the reward for monogamy is pleasure. In the ultimate sense, the reward is reproductive success.

This finding also helps to explain why monogamy seems to be easy for mammals to evolve into: the brain circuitry is already there. Flipping a switch or two is all it takes.

Monkey Business

Primates are the order of mammals to which humans belong and have more monogamous species than any other order, around 15 percent. The first primates emerged around the end of the age of dinosaurs and they eventually diverged into four major groups: the prosimians (lemurs, bushbabies, etc.), the Old World monkeys (baboons, macaques, and others native to Africa and Asia), the New World monkeys (various monkeys native to Central and South America), and the apes. Primates tend to be forest dwelling and most are arboreal, using their opposable thumbs for ably grasping branches. They are also known for their relatively large brain, good visual acuity and color vision, and complex social systems. Before getting to our own group, the apes, there are a couple monkey species worth exploring.

The coppery titi monkeys (*Plecturocebus cupreus*) are often referred to as the primate with a family structure closest to our own. One male and one female form a group with their children, a nuclear family, which can occasionally incorporate a third generation as well. It was long known that these monkeys form strong heterosexual pair-bonds and mate for life, but in 2020 it was discovered that they are also genetically monogamous as well.[35] Thus, these monkeys have the distinction of being the eighth mammal species discovered to exhibit true sexual monogamy. That's just eight out of more than five thousand species, or 0.0015 percent.

There is almost no dimorphism in coppery titis: the males and females look identical and behave very similarly. They partner very well, sharing as much of the child-rearing duties as possible. While females do the nursing (duh!), the fathers do most of the work carrying the children around before they can walk. They also work equally to procure food and defend territory and family. The attachment between mated pairs of coppery titis is very intense, and partners become distressed in each other's absence. Both sexes are possessive and exhibit mate-guarding behaviors around other adult titis. In other words, they are extremely jealous. In fact, immediately after it was discovered that they are sexually monogamous, they became the focus of research on the neurobiology of jealousy, since monogamy and jealousy go hand in hand.[36, 37]

The same study that demonstrated sexual monogamy in coppery titis also revealed a key reason why sexual monogamy is tenable for them: mated pairs are not close relatives, which means this species already has a system in place to prevent inbreeding, the big danger that promiscuity is aimed to prevent. In titis, both males and females instinctively leave their home range when they mature into adulthood. This is called *natal dispersal*, and with both sexes doing it, and the titi population being large enough, it appears to be sufficient to ensure a genetically robust population. This is another example of how males and females of this species are wired for the same behaviors.

Males and females being on equal footing appears to be a key ingredient in the recipe for monogamy. Of the eight sexually monogamous mammals that have been discovered, three are primates. All three—Azara's night monkeys, Müller's gibbons, and the coppery titis—exhibit essentially no sexual dimorphism: the males and females are physically indistinguishable, behave similarly, and equally contribute to parental care of the offspring.[38] This is not common. In most primates, females and males have different markings or different body sizes. Gorilla males are almost twice as large as females, for example. And behaviors are usually sexually dimorphic as well, given the realities of pregnancy, nursing, and the long developmental period that primate offspring need. The fact that the three primates found to be sexually monogamous are also sexually monomorphic can hardly be a coincidence. There are

The cover of Not So Different *features a "family" of Barbary macaques.*

many primate species that are monomorphic without being sexually monogamous, so this is clearly not enough, but it may be a necessary starting point for sexual monogamy to emerge. And this makes sense because monogamy is a type of equalizer between the sexes.

And now we turn to a monkey species so dear to me that I put them on the cover of my first book, *Not So Different: Finding Human Nature in Animals.* There is no shortage of fascinating facts that make Barbary macaques (*Macaca sylvanus*) unique.[39] They stem from the phylogenetic root of all macaques, meaning they are the oldest branch from the ancestors of all twenty-three macaque species. They are the only macaques found outside of Asia, and the only nonhuman primate to live north of the Sahara Desert. These resourceful midsize monkeys also have the unique distinction of being the only nonhuman primate to have successfully colonized Europe in the recent past: stowaways on ships from Morocco to Gibraltar have established a stable population of more than three hundred individuals.

Germane to this book, Barbary macaques are unique among primates for their system of distributed alloparenting.[40] As mentioned above in the discussion of lions, alloparenting is when groups of animals contribute communally to the care of offspring, regardless of parentage. The lionesses of a pride will care for each other's children, even allonursing them. But Barbary macaques go even further because males and females contribute equally to the alloparenting of all young. Male Barbary macaques are constantly carrying infants around, grooming and playing with them, and helping them eat once they are weaned.[41] Even the highly egalitarian bonobos (next section) do not show this degree of distributed parenting.

Barbary macaque troops are also matriarchal, with dominance determined by relatedness to the top-ranking female.[42] Males form coalitions, or friendships, and are often invited into social interactions by one male handing another male an infant to care for![43] Although they engage in sex more often during estrus, females enjoy sex throughout the year, with both males and females. The males are also highly promiscuous and bisexual in this very sexual species, but heterosexual sexual access (mate choice) is female-driven and mostly follows the dominance hierarchy, meaning that females prefer males of high social status.[44] Because of all the sex going on, paternity is a complete mystery. And the mothers only weakly favor their own genetic offspring.

The distributed alloparenting of Barbary macaques is truly impressive, and it is also unexpected. Among mammals, when paternity is obscured, we generally see *less* paternal investment, not more. In the mind of a mammal, if they don't know which child is theirs, they usually don't bother investing in any. But in these macaques, it seems to be the opposite. Since any of the children could be theirs, they invest in them all. And what's more, the communal parenting seems to be the social glue that holds the group together. Could this be a result of the matriarchal structure? It's possible that it could be a contributing or necessary factor, but it's definitely not enough, since there are many matriarchal species and only the Barbaries are this special.

While there are no mated pair-bonds in Barbary macaques, they do show strong social bonding and attachments. In fact, social partners, particularly males, can form short-lived dyads while caring for youngsters together, like little temporary families that form for the purpose of raising children.[45] This is why I put that photo on the cover of my book. What looks like a traditional family of mother, father, and baby is actually a male-male dyad caring for a child that is almost certainly not the genetic offspring of either. As my husband and I are currently raising two children that are not our genetic off-spring, this photo resonated so strongly with me that I also memorialized it with a tattoo on my shoulder blade.

Due to their egalitarian nature, highly invested fatherhood, and matriarchal structure, we might be tempted to expect little or no sexual dimorphism in Barbary macaques, like in the titi monkeys, but we would be wrong. This is because sexual monomorphism seems to correlate with monogamy, and the Barbary macaques are the opposite of monogamous. They don't form dyads, and everyone has sex with everyone else.[46] In terms of dimorphism, males are about 15 percent larger than females, and females have genital swellings that make them easy to spot. They behave differently also, forming sex-segregated coalitions and a female-led dominance hierarchy for grooming and food sharing. Females care for the youngest infants, but then males take over most of the parenting duties as soon as the infants start to wean and are more physically capable.

The titi monkeys, with their total lack of sexual dimorphism and strong monogamy, seem like the opposite end of the spectrum, as Barbary macaques are sexually dimorphic and show no monogamy, but they actually have some things in common that put them at odds with most other primates. First, they both show high paternal investment in offspring.* And second, both titis and Barbaries exhibit a high degree of female mate choice and overall autonomy. In many mammals, males are dominant, aggressive, brutish, sexually controlling, and coercive, while females have little choice but to endure it. In these two monkey species, however, females are fully empowered.

At the risk of romanticizing, both coppery titis and Barbary macaques have both gender equality and social groups that live in relative harmony. With titis, the groups are nuclear families, while in Barbaries, they are large blended families, but in both, the members work together cooperatively and mostly free of sexual conflict. This shows that there are many paths to a peace-

---

* In more than 85 percent of primates, and 90–95 percent of mammals overall, fathers show little or no paternal care for offspring. The internal gestation of mammals really sets females up to get short-changed in the parenting department. Males can simply drop their seed and go, leaving females with the lion's share—or shall we say the *lioness's share*—of the work.

ful existence, and both monogamy and promiscuity can get you there. The common factor, at least in these two primate species, is the empowerment of females.* When males are in charge, as in a harem, life can be pretty rough—especially for the males! But when females are in charge, harmony and equity seem to be the natural results.** It's something to ponder.

Aping Around

Let us now inch even closer to our own species. Humans are in the group of tail-less primates known as apes. Our closest living relatives are the two sister species of chimpanzee, the common chimp and the pygmy chimp, also known as the bonobo. Our next closest relative is the gorilla, and collectively, chimps, gorillas, and humans are known as the *African apes*, having spent all or nearly all of the last 10 million years of evolutionary history on that continent. Going further out, our next closest relative is the orangutan, and collectively, all of these species are called the *great apes*, for virtue of being larger than the remaining species of ape, the gibbons, sometimes called the *lesser apes*. Orangutans and gibbons have spent all or most of the last 10 million years in Asia and so are referred to as the *Asian apes*.

In our effort to discover the natural state of human sexuality and sexual relationships, it makes sense to focus on our closest relatives, the ones with whom we share the most recent common ancestor. We share a common ancestor with chimpanzees that lived about 6 million years ago. The common ancestor of chimps, gorillas, and us lived about 10 million years ago, and the common ancestor of all great apes lived about 12 million years ago, give or take. That's about as far back as we will go because, as we will soon see, just that recent history provides an astonishing amount of sexual diversity.

---

* Keep this fact in mind in chapter 7, when I discuss matriarchal human societies.

** An important exception to this are hyenas, which are matriarchal but also marked by fierce competition, hostile dominance, and violent conflict. However, hyenas have levels of circulating testosterone that are so high that even female bodies are markedly masculinized. In a sense, they are the exception that proves the rule.

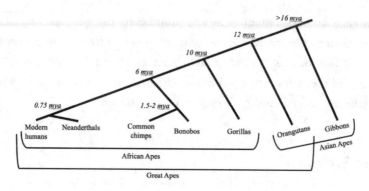

Evolutionary relationships among living apes (mya, millions of years ago)

Let's begin with orangutans, our most distantly related fellow great ape.[47] There are two species of orangutan,* one native to Borneo (*Pongo pygmaeus*) and the other to Sumatra (*Pongo abelii*), and they are the most solitary of all apes. Adult males are almost completely solitary. As they approach adulthood, they become transient, roaming on their own for a while as they reach their full size and strength. Orang males are territorial, establishing a home range that they exclusively dominate, with frequent skirmishes as they jockey for the largest and most enviable range that they can defend. Challenges are frequent, and most males take over another male's territory rather than stake out a new one. It appears that adult male orangutans do not form social bonds of any form.

Female orangutans are a bit more social. They also establish home ranges but are friendly with neighboring females and generally tolerate overlaps in home ranges. In fact, females generally do not venture far from their natal range, so neighboring females are often sisters or otherwise close relatives. Because females do not roam, it is important that males do to prevent inbreeding.

When it comes time for breeding, males attempt to attract ovulating females through a combination of physical display—long hair and flanged

---

* A tiny isolated population of orangutans in Northern Sumatra, which was previously considered a subspecies, was recently reclassified as a third species, the Tapanuli orangutan (*Pongo tapanuliensis*), based on genomic analyses that revealed their genetic distance from the other two populations. Besides making louder and higher-pitched calls, there are no major differences in sexual behaviors.

cheeks—and vocal calls. Males often attempt to force copulation with females they encounter, whether they are fertile or not. The females tend not to resist copulation if they are not ovulating. Since they are unlikely to become pregnant anyway, there is no need to be choosy, and I suppose it's just not worth the fight to resist. When they are in estrus, however, females are much choosier and males must chase, subdue, and overcome substantial resistance in order to successfully copulate with them. This resistance to copulation by a female in heat is believed to be a test of quality control that only healthy, strong, and virile males can pass.

When a male orangutan succeeds in subduing an ovulating female, he copulates with her repeatedly, almost constantly, over a period of weeks until he is sure that she is pregnant. He jealously watches and restricts her every movement during this time, barely eating or sleeping. This is mate-guarding in the extreme. And what happens next is even worse: once he is sure the female is pregnant, the male bolts and never returns. Orangutan fathers are true deadbeats: they force themselves on the females, engage in mate-guarding to protect their paternity, and then they abandon the pregnant female in order to seek out another.

Because pair-bonding does not exist and social bonding is weak, there is little we can learn from orangutans when it comes to human sexual behaviors. But I think they are worth discussing if only to point out how quickly different social behaviors can evolve, as these apes approach sexual relationships entirely differently from all of their closest relatives.

As discussed in chapter 2, sexual behaviors in gorillas* are strikingly different from that of orangutans but some parallels are indeed noteworthy, especially the tyrannical nature of the adult males.[48] Gorillas construct the harem lifestyle,** usually with a single dominant male (the silverback) and several

---

* Gorillas were thought to be comprised of one species, with three local variants as subspecies. They have since been reclassified as two species, *Gorilla gorilla* and *Gorilla beringei*, with two subspecies each. These are separated geographically, but their social behaviors are similar enough to be discussed here as one species, for simplicity.

** The exceptions to this are one population of mountain gorillas living in Virunga National Park in Rwanda, which, as described in the final section of chapter 2, have abandoned the harem lifestyle and have been living in large multi-male, multi-female groups for three decades now.

females. As in lions, the males are less involved in the social life of the group and, instead, the females form a dominance hierarchy with an alpha female who is, in many ways, the social leader of the group. She keeps the others in line, resolves conflicts, and controls social access to the silverback. Although the silverback has unfettered *sexual* access to all of the females, the females jockey for social position within the group, which comes with feeding priority and other privileges.

Unlike in lions, the female gorillas in a harem are not all related to one another, but most harems will include some sisters. These sisters are even more closely bonded to each other, and they use these sibling alliances to advance their position in the group. There is a great deal of rivalry and politics in a harem, but the social bonding among the members is tight. Unlike in orangutans, the males are also part of this social bonding and form very strong attachments to their wives and their children. For example, the sudden death of a group member leads to turmoil and visible anguish in the surviving members, including the silverback, which gives way to grief and eventually social restructuring. Social attachments are strong in gorillas, including and especially the pair-bonds between the silverback and the females. They treat each other with care, compassion, and affection.

The harem lifestyle is much harder on male gorillas than it is on females. As a young male approaches sexual maturity, he becomes a threat to the silverback, who is his father. This is quite a switch because gorillas are doting fathers, especially to their sons. The mothers are also quite chauvinistic, in fact, favoring sons over daughters for food sharing, grooming, and other social privileges. This is easier to accept when we remember that the females have a much more assured path to success and a life of ease than do the males. As with lions, males must eventually leave the group, enter a solitary period as they reach their full size, and then attempt to take over a group either by challenging an established silverback or recruiting younger adult females to form a new group.

When a silverback is eventually deposed by a younger, stronger male, as with most other harems, all of the young gorillas are slaughtered by the new silverback. This brings the females into estrus and diverts their maternal investment into his children, rather than the previous alpha male's. This is also why young males never challenge their own father. In a sense, their genes are already enjoying success through their siblings and half-siblings, and to kill all of the youngsters would be self-defeating from an evolutionary point of view.

In sum, while gorillas are not monogamous, they do experience strong social bonding and attachments, alloparenting, and group cooperation and affiliation. Aside from the occasional (and very dramatic) lethal male-male contests, skirmishes are rare.

All in the Family

And now we shall turn our attention to our very closest living relatives: the chimpanzees. As good luck would have it, there are two closely related sister species of chimpanzee: the common chimpanzees (*Pan troglodytes*) and the pygmy chimpanzees, also called the bonobos (*Pan paniscus*). I call this good luck because these two species are closely related and very similar in most ways, but they are *very* different when it comes to gendered behaviors, making them highly instructive in our tour through the sexual behaviors of our fellow apes.[49]

Although both chimp species are highly threatened by poaching and habitat loss, common chimpanzees are much more numerous and widespread than the bonobos, which are limited to a small region in the Congo. The two species are believed to have diverged upon the formation of the Congo River less than 2 million years ago. Because chimpanzees are not good swimmers, the river fragmented the ancestral chimpanzee population in two. The population that was isolated south of the river became the bonobos of today, while north of the river, the larger population was spread more broadly across central Africa and became the common chimpanzees.

First, their similarities. Both species live in highly social, highly complex, and highly dynamic social environments. They both create large communities of animals, up to 100–150 affiliated animals called *troops*, but they break down into many small groups for day-to-day living. These small groups, sometimes called *parties*, are usually less than ten individuals and can be all-male, all-female plus their offspring, mixed adult animals, or even a small family of a mother and her children. Chimpanzee troops are called a *fission-fusion society* because members can move among different small groups, and the groups can form and dissolve spontaneously. Some groups exist for only a quick purpose, such as a hunting raid, while other groups persist for months, such as a female with a nursing child and her other dependent children. The larger communities are affiliative and gregarious, but members of one troop are hostile to outsiders from other troops. Violent conflicts between troops are not infrequent, and common chimpanzees are known to organize murderous raids against other troops.

In terms of social behaviors, this is where the similarities between the two species end. The common chimpanzees are patriarchal and employ a very strict and linear dominance hierarchy.[50] There is a male rank order and a female rank order, and all males are dominant over all females. The dominance order is established and maintained through aggression, especially among the males. While threats and displays are the preferred method of establishing dominance, combative challenges are not uncommon either and can sometimes be lethal. Males are especially hostile to males from other troops, usually attacking them on sight. When a dominant male returns from foraging, whether alone or with others, his first order of business is to aggressively remind everyone that he is in charge.

Common chimpanzees are also highly territorial, each troop maintaining and defending a large home range filled with fruits, seeds, leaf buds, and other foods they prefer. The borders of the territory are patrolled by roaming parties of high-ranking males, with the main adversaries being chimpanzees from rival troops. These roaming parties will also hunt small rodents and monkeys on occasion, and so the chimpanzee diet does incorporate some meat and insects in addition to the vegetation on which they mainly subside. Males do not leave the troop into which they were born, instead jockeying for rank, either through direct combat or by forming temporary alliances with other males. This means that the males of a given troop are almost always close relatives of one another. While this does facilitate some coalition building, the alliances are fickle and self-serving, males are exploitative and manipulative, and male-male aggression, fighting, and lethal violence are a fact of life in chimpanzee society.[51]

Female common chimpanzees, on the other hand, almost always leave their natal group once they reach sexual maturity and join another community. This promotes genetic diversity and reduces inbreeding, crucial concerns in chimpanzee societies since the males do not migrate. Usually, when a female joins another troop, she remains there permanently, but occasionally she returns to her natal group after becoming pregnant. Both of these outcomes accomplish the same task of exchanging genes between the two communities and enriching the gene pool.

The instances of females returning to their natal groups while carrying a pregnancy from a foreign father raises an interesting question: Where will the paternal investment and protection come from? Will this fatherless child be at danger of infanticide, like a gorilla child would be? The answer is that chimpanzee fathers do not engage in childcare anyway, so paternal investment is moot, and while infanticide does occasionally occur, the offspring of foreigners are not at increased danger. The reason is that chimpanzees are prolifically promiscuous, to the point that no one *ever* knows who anyone's father is.

When a female chimpanzee is in heat, her genitals engorge with blood, advertising her fertility. She then pursues a promiscuous mating strategy, having sex with several males, and she can pursue a variety of options. Sometimes the dominant male of a group will control access to fertile females, giving himself the most access and granting limited access to his allies. Other times, a male and female will engage in a private courtship away from the rest of the community. She may even go to another troop, mate with whatever males she finds there, and then return. In all cases, females mate with enough males to obscure paternity, apparently by design.

The result of this is that the male reproductive strategy in chimpanzees is to have sex with as many fertile females as possible, as many times as possible, in the hopes that one's sperm will outrun and outnumber the rest. This is called *sperm competition* and leads to the evolution of exceptionally fast and numerous sperm cells, voluminous ejaculate, and obnoxiously large testes. In terms of overall body size, chimpanzees are 10–20 percent smaller than humans, but their testes are *three times larger* on average. That's how strong a selective pressure sperm competition is. Indeed, it is a strong general trend in birds and mammals, including and especially primates, that when the females are promiscuous, the males have big balls.[52]

Gorillas, by the way, are at the opposite extreme. Because polygyny does not allow for female promiscuity, sperm competition is negligible. Accordingly, males have tiny testes (and even tinier penises).* Where humans fit in this scheme has been a matter of debate for decades. On the one hand, our testes (per body weight) are much bigger than those of the polygynous gorillas, but much smaller than those of the promiscuous chimps. This has confounded arguments on both sides of the debate about whether monogamy is humanity's natural state.

With no one knowing which offspring are from which father, chimpanzees engage in no direct paternal investment. However, male chimpanzees do offer general protection and food procurement for all the youngsters in their troop. They may not know which child is "theirs," if any, but they do behave as though they recognize that all or most of the children are their relatives, nieces and nephews, cousins, half-siblings, etc. Even if a female is impregnated by a foreigner and returns to her natal troop, she herself is likely related to the males in the troop. And when a female does join another troop, she may not be related to the males, but her children will be. Promiscuity sets up a system of distributed paternity that roughly follows the dominance hierarchy. The most successful males father the most offspring, but all of them look after the juveniles to some degree.**

For the females, it's a different story. Since they know who their children are, they bond with them tightly, and chimpanzee infants are entirely dependent upon their mothers. They must be carried by their mothers for the first six months; then they ride on their backs for another eighteen months before they are finally able to amble about at around two years of age. Mother and

---

* For this reason, the multi-male gorilla groups described in chapter 3 are believed to be a new and rare social arrangement. We would predict that the resulting female promiscuity would lead to the evolution of large testicles, but, oddly, females in the multi-male groups appear to have sex much less frequently than do harem females.

** Once again, humans are a mixed bag. Human fathers preferentially dote on their own biological children, like gorilla fathers do, and are prone to mate-guarding (jealousy). But, as in chimpanzees, humans always live in multi-male groups with strong male-male social bonding, and the concealed nature of human ovulation means that paternity cannot be known for certain.

child are in constant physical contact for the first two years, and children are not fully weaned until they are five years old. Only by ten years will juveniles begin to move between groups on their own and become independent. Yes, chimpanzee mothers spend ten full years dutifully raising their offspring, nursing and feeding them, teaching them how to build nests, find food, play with other chimps, and all the other things that chimpanzees need to know, with only occasional help from other males and females.

This is all to say that, while chimpanzees don't pair-bond, they most definitely form very strong social and family attachments. Besides the unmatched mother-offspring bond, chimpanzees form attachments to their siblings and their affiliates. There is also social bonding through the dominance hierarchy, and almost everyone experiences a level of attachment to the highest-ranking males and females. Chimpanzees grieve the loss of their attachments much like we do: Jane Goodall describes a chimpanzee who essentially died from grief following the death of his mother. Social bonds are strengthened through grooming (picking lice and insects out of each other's hair), play, eating together, and other friendly contact.

This indicates that the oxytocin-attachment system is very strong in this species, even though it doesn't seem to get activated by sexual activity.[53] While this may seem odd to us, it makes perfect sense for how chimpanzee society is structured. The bonds that are important for survival and success are indeed strong. But pair-bonding between sex partners—in other words, monogamy—just doesn't benefit anyone, so it's absent. This is in contrast to gorillas, who, despite being polygynous, form strong attachments between sex partners. The silverback is strongly bonded to the females and they to him. As any parent with more than one child can attest, pair-bonds do not have to be in a 1:1 ratio to be very strong.

Surprising though it seems, the primate species that is the most different from the common chimpanzee when it comes to sociosexual behaviors is also its closest relative: the bonobo. These two sister-species share over 99 percent of their DNA sequences and have nearly identical diets, natural lifespans, physiology and health concerns, and so on. And yet they have evolved radically different, almost polar opposite, social structures. Where chimpanzees are aggressive and domineering, bonobos are friendly and sociable. Where chimpanzees are male-dominated and hierarchical, bonobos are matriarchal and egalitarian. Where chimpanzees are territorial and vigorously defend a home range, often with lethal effect, bonobos are more nomadic, and encounters with strangers rarely result in conflict.

But nowhere are the differences between chimpanzees and bonobos more stark than sexual behaviors.[54] Bonobos are the sluttiest creatures ever described. Sex is used for greeting, affiliation, conflict resolution, reconciliation, to express excitement, to facilitate food sharing, and also, well, just because. Bonobos are constantly having sex and display no aversion based on sex, age, or relatedness (with the exception of mothers and their adult sons). Sex between adults and children, even infants, is frequent, although usually not penetrative, and children initiate sexual contact as much as adults do.

A good example for just how different these two species behave is what happens when they encounter bountiful food.[55] When common chimpanzees happen upon a large food source, access is determined by dominance status and enforced by aggressive posturing or combat, if necessary. When bonobos happen upon a food source, they break into a frenzied orgy. It is believed that the group sexual activity reduces tension, relieves excited agitation, and promotes harmonious food sharing.

Bonobos also get very creative with their sex.[56] They engage in tongue-kissing, oral sex, and mutual masturbation, and they are the only primate other than humans to have sex in the face-to-face position. Male-male sex includes fellatio, anal intercourse, frottage, and penis-fencing, while female-female pairs engage in cunnilingus, manual stimulation, and genito-genital rubbing, as discussed in chapter 4. Bonobo females go into estrus much more frequently than do common chimpanzees, despite having very similar birth rates, gestation time, and interbirth interval (time between pregnancies).

While common chimps are patriarchal, bonobo society is said to be matriarchal because the most dominant position is always held by an alpha female. However, dominance is won not through combat or aggression, but through age, alliance building, and sexual activity. Bonobo females can "sleep their way to the top," so to speak, and males can gain a high station through their affiliation with high-ranking females, especially their mothers.[57] That said, the dominance hierarchy itself is weak, and sexual access is mostly granted all around. As in common chimpanzees, mother-offspring bonds are particularly strong, but all members of a bonobo troop work together to protect and care for younglings.

In addition to having sex many times every day, bonobos spend a lot of time playing, even as adults. Among juveniles, playing together builds trust and social bonds, and adult females show a strong preference for males that they often played with as children. While in common chimps, aggression is how dominant males obtain and maintain sexual access to females, in bonobos, male aggression is a major turnoff to females. For a male, the best strategy for mating access to the highest-ranking females is by earning their trust, preferably when they are young, or that of their friends. In bonobo social life, friendships and alliances are everything.

How these two species evolved such different social behaviors despite having almost indistinguishable genetics and living in nearly identical habitats is something of a mystery, but the most accepted theory holds that the tiny population that found itself stranded south of the Congo River enjoyed nearly unlimited access to the preferred foods of ripe fruit, young leaves, insects, etc., so there was no food competition between troops and no need to establish and defend a home range. So territoriality and aggression toward outsiders offered no benefit and slowly disappeared. In addition, the southern population was initially small and remained so due to the low birth rates and long generation times of great apes. With ample food and space, and negligible competition for resources or mates, a harmonious, egalitarian, sex-crazed social environment was free to emerge.

There has been much speculation as to what we can learn from the behaviors of our two very different closest relatives, and what lessons, if any, we can apply to our own behavior.[58] Until recently, most anthropologists and evolutionary psychologists attempted to draw comparisons between humans and common chimpanzees, but never to bonobos.

In fact, when one of the most influential scientists in the field of human evolution, Louis Leakey, decided that the best way to understand how human behavior evolved was to understand how the great apes behave, he hand-picked three scientists to study orangutans, gorillas, and common chimpanzees. Those three scientists, Biruté (Mary) Galdikas, Dian Fossey, and Jane Goodall, respectively, came to be known as "The Trimates," and their pioneering work laid

the foundation for not just our understanding of ape behavior but how to actually study apes in the field. We knew almost nothing about ape behavior before
these women did their work, and their methods gave rise to the modern field
of Western primatology.* But because bonobos were not covered by the foundational efforts of "Leakey's Angels," they were largely ignored for decades.

Because of the focus on common chimps, male dominance came to be
understood as a natural and expected part of the human condition. As the
mostly male scientists through the last two centuries looked around at the animal world, they were all too eager to find validation for male dominance in
human society, including and especially in the sciences. And the communities
formed by common chimpanzees provided that validation, despite Goodall's
persistent efforts to highlight the crucial role that females play in structuring
that society. In the common chimps, there is no sugarcoating that males dominate and they do so through strength, threats, and combat when necessary.
And there are certainly parallels to various aspects of human societies around
the world.

But what about the bonobos? We are just as related to them as we are
to common chimps. The two are close sister species, and both are our cousins from the same branch of the evolutionary tree. While some scientists and
commentators see parallels to human society in the way that chimps behave,
there are also strong parallels to bonobos.

For example, humans and bonobos are both sex-crazed. If bonobos didn't
exist, humans would easily be the most sex-crazed primate species. While the
females of both chimp species advertise their fertility with swollen genitals,
common chimps concentrate their sexual activity during the fertile period.
Bonobos have just as much sex when they are not ovulating as they do when
they are. Human women also have sex throughout their menstrual cycle, and
while already pregnant or nursing. In fact, humans have *concealed* ovulation,

---

* We call the field *Western* primatology because, by this time, scientists in Japan had been
studying the social behaviors of native monkeys in their natural habitat for decades. This meth-
odological tradition is so developed that "Japanese primatology" is a scientific discipline with
its own history, methodology, and body of knowledge. Currently, Western and Japanese prima-
tologists are in the process of integrating the knowledge sets of the two fields, while maintaining them as distinct and complementary traditions.

where no one knows who is fertile or when, partly in order to promote constant and frequent sexual activity, thereby reinforcing the pair-bond.

In addition, bonobos and humans are the only primates to have sex face-to-face and to spend so much time in sex acts that couldn't possibly lead to conception, including oral and anal sex, kissing, manual stimulation, mutual masturbation, and so forth. And speaking of time spent having sex, the average time for a copulation event in common chimps is *seven seconds*. Whereas humans and bonobos both like to take their time and pursue long sexual sessions purely for the pleasure of it.

And the parallels between humans and bonobos go beyond the sexual realm. While yes, human society is often violent and marked by a might-makes-right scheme of dominance, we are also a species marked by peaceful conflict resolution. The odds of a common chimpanzee dying through violent conflict with another chimpanzee is somewhere around a hundred times greater than for bonobos or humans.[59] Furthermore, while human societies are territorial, like common chimps, borders are most often established and maintained through agreements and mutual nonaggression, rather than armed conflict, even in traditional hunter-gatherer societies.

In fact, the way that different societies interact is another example of human-bonobo similarity. In common chimpanzees, two unacquainted individuals—that is, two individuals from different troops—will be violently aggressive toward each other on sight. Strangers are always viewed with suspicion and met with threats of lethal violence. But this is not the case for humans, is it? My friend Mark Moffett, the world's expert on animal societies, uses the example of a human casually strolling into a coffee shop surrounded by perfect strangers.[60] If a scene like that played out with chimpanzees, that stranger would not leave the coffee shop unscathed. He probably wouldn't leave the coffee shop at all. But in human societies, despite our xenophobia and racism, we are generally tolerant of nonthreatening strangers. This too mirrors the bonobo approach to strangers: they may be wary, but they don't resort to violence. Usually, to ease the tension, they simply have sex with the stranger. If this doesn't sound like a human strategy to you, then you haven't spent much time in hotel bars.

Personally, I think the real lesson on human nature is not in figuring out which chimp species we most closely mirror, but rather in the interesting contrast between the two chimp species themselves. We are separated from the chimps by at least 6 million years of distinct evolutionary history, but the two chimp species are separated from each other by only about 1.5 million years. Apparently, that is plenty of time for all three species to have developed their unique approaches to sex and society.

It may sound like a very long time, but by 1.5 million years ago, *Homo erectus* had already spread through the Middle East, Central and East Asia, and several Pacific islands. And *Homo sapiens* was just about to burst onto the scene. In the grand scheme, it's not that long ago. The fact that the two chimp species diverged so starkly in their socio-sexual behaviors, and in such little time, demonstrates how quickly and easily these behaviors can be molded. There is an inherent plasticity to the way that apes pursue sexual activity and for what purposes. By now you may be getting tired of me repeating this refrain, but it seems as though sexuality is an aspect of our nature that is especially prone to diversity, flexibility, and experimentation.

The conundrum we face, though, when examining the sexual similarities we have with our closest relatives is our relationship to monogamy. Neither species of chimp is socially or sexually monogamous, even a little bit. They do not form pair-bonds with sexual partners at all, while humans clearly do. Our oxytocin/vasopressin attachment system is activated in a unique way by sexual activity. That intense feeling of being "in love" is clearly reserved for current or prospective sexual partners. In this way, we are unlike either chimp species and more like the gorillas who, despite being polygynous, form strong attachments between sexual partners with all of the associated emotions, including longing, jealousy, and profound grief when an attachment is lost.

It is worth noting that the oxytocin attachment system is still present and very strong in both chimp species. They form strong mother-child bonds, as well as social bonds—friendships—that are enduring and profound. The attachment system just isn't activated by sex. So among the various species of apes, all of us form strong familiar and social attachments, but the way that sex activates these attachments, or doesn't, is different.

Although not directly related to monogamy, researchers have found two genetic similarities between humans and bonobos that may provide a common basis for the lover-not-fighter differences that we have with common chimpanzees. First, chimpanzees have a slightly different version of the oxytocin receptor gene that makes their receptor less sensitive to oxytocin. Fascinatingly, there is a different minor variant in the oxytocin gene found in a small minority of humans that also renders the receptor less sensitive, and individuals who harbor this variant have deficits in empathy and prosocial behavior (like common chimpanzees do, in my opinion). Second, in 2014, researchers found that common chimps also have slightly different versions of the vasopressin receptor than do humans and bonobos.[61] Once again, there are variants in this gene among humans that correlate with levels of social bonding. While these correlations are far from direct proof, they are highly suggestive that behavioral neurocircuitry involving oxytocin and vasopressin may be key to how humans, chimpanzees, and bonobos evolved their unique sociality.

What conclusion can we draw from our exploration of monogamy in the animal world? First and foremost, we have confirmed that social and sexual monogamy are altogether different things and that social monogamy does not imply sexual monogamy. Second, we must concede that sexual monogamy is vanishingly rare among mammals, is usually temporary when it is observed, and is not found in any ape species for any length of time. Third, we know that attachments and pair-bonding between sexual partners is highly variable throughout mammals and that every species approaches this differently, if at all. And fourth, pair-bonding itself is present among apes but not common, and when it is observed, it does not involve 1:1 sexual monogamy.

In other words, every species, and even different populations within species, approaches sexual relationships differently and in a way that makes the most sense for the social environment in which they exist. The striking degree of sexual diversity, even among close relatives, indicates that these behaviors are highly flexible due to the vagueness of the sex drive and the fluidity of attachment and bonding. This diversity makes it difficult to draw any broad conclusions about sex, except for one: sexual monogamy is exceedingly rare.

Chapter 5
Sexual Animals

The notion that the sole purpose of sex is for the conception of offspring is one of the most widespread misconceptions in all of biology, but of course it is understandable. The creation of new life is definitely a momentous consequence of sex, and it is also an outcome that cannot come about any other way (until the advent of assistive reproductive technology). And this is also the *foundational* purpose of sex—the reason it evolved in the first place, and the common purpose that animal sexual function shares with other kinds of organisms such as plants, fungi, and microorganisms. So yes, a chief function of sex is to make genetically unique babies, but that is just the beginning. Animals have taken this intimate act and done so much more with it.

The real genius of sex is that it feels so good. Animals have a sex drive, an instinctual urge to pursue sex according to the particular attractions that the animal experiences. The sexual appetite is self-reinforcing because of the way that sexual activity stimulates the reward center in the brain. When we satisfy our urge to have sex, we experience the sensation we call pleasure. Pleasure is one of those truly ingenious inventions of evolution because it can be used to drive a variety of behaviors.[1]

The reward center is not just activated when we have sex but any time that we satisfy an appetite or drive. When we are dehydrated or overheated, drinking a cold glass of water stimulates our reward center. When we eat delicious food (or any food when we are hungry enough), we experience pleasure. It feels *so good* to scratch an itch. The thrill we experience during intense sporting or video games, lying down when we're tired, a hug when we're lonely, even a cold breeze on a hot day—these all stimulate the reward pathways in our brain, a sensation that then trains us exactly how to satisfy these various drives and instincts.

For the purposes of this discussion, we don't need to cover all that is known about how pleasure and reward are constructed by our brains. Suffice it to say that it is quite complicated, and many brain areas and pathways cooperate. The two neurotransmitters that are common to all forms of pleasure are *dopamine* and *serotonin*, with the former acting on shorter timescales and the latter on longer ones, but both work together to create the sensation of pleasure and a sense of well-being. The different flavors of pleasure involve additional neurotransmitters, neural pathways, and brain centers, but dopamine and serotonin are central to all forms of pleasure. In fact, the addictive nature of drugs such as cocaine and the opioids stems from their direct activation of dopamine-responsive pathways in our reward center. And this is why we can also get addicted to behaviors as well as substances, such as gambling, stealing, exercising, eating, and, of course, sex.

Besides dopamine and serotonin, sexual pleasure also involves acetylcholine, epinephrine (adrenaline), norepinephrine, oxytocin, vasopressin, and other neuropeptides; and many areas throughout the brain are activated.[2] Because cultural taboos have prevented this area of our biology from receiving much grant funding for scientific research, we don't know as much about sexual pleasure as we do hunger, thirst, and even drug addiction. But what we do know is that, as mysterious and otherworldly as it may feel, sexual pleasure is the result of nerve impulses and chemical reactions in our brain.

In her infinite wisdom, nature has linked sex with pleasure as the way to drive us to pursue this behavior, much like hunger drives us to pursue and consume food. In whatever animal population this linkage first evolved, the experience of pleasure drove them to have sex and reproduce prolifically, which gave them an advantage in the great competition of natural selection, and the evolutionary cycle continued. Behaviors that lead to success proliferate, while those that don't eventually disappear. Once sex and pleasure were linked, the resulting success created a cycle that strengthened that linkage, but it also created opportunities for new inputs. Since sex feels good, animals will try to have a lot of it. Those that do will have more offspring. Those offspring will also experience sexual pleasure, and so on.

But the pursuit of sex must be balanced with the many other pressures that animals experience in the wild. Consider a population of animals that are already reproducing about as fast as they can, given the carrying capacity of their environment (availability of food, space, other materials and resources). Having more and more sex wouldn't help, since producing more offspring that will simply compete with each other won't expand the population. But having more sex *could* give one animal an edge over another. If the more sexually prolific animals have more offspring in the next generation, this would lead the species toward more sexual activity (think bonobos). But pulling in the other direction are all the other needs of the animals. If an animal too often pursues sex instead of food, he may starve or become undernourished enough to reduce his ability to attract sexual partners. The various needs and pressures will eventually find some kind of balance, an equilibrium point, in which the drive for sex competes with all of the other drives an animal experiences, including their need for rest and, in social species, allies. It can't be all sex, all the time.

It also should not surprise us that sex performs many different functions in animal societies. In evolutionary terms, behavior is not all that different from anatomy. It's influenced by genetics but takes shape in an environment, and it is subject to the same evolutionary forces, random variation followed by natural selection, and all that. And just like anatomy that originally evolved for one function can later evolve for another, so can behaviors. Just as feathers originally evolved for warmth but later became crucial for flight, sex first appeared to bring together sperm and eggs but now can be used to strengthen pair-bonds. This is called *exaptation* and is a major theme in evolutionary biology. From molecules to organs to behaviors, life constantly comes up with creative new uses for old things.

What do we suppose happens when having sex brings some additional benefit for the animal? This would be favored as well, as evolution rewards success. Let's first consider a nonsexual example. Young kittens (both wild and domestic) can often be seen "hunting" small bugs and even inanimate objects as a form of play. This behavior is clearly instinctual, but what does the kitten get out of it? Why has this instinct evolved? The reason is that play hunting is a great way for kittens to exercise and train their bodies for how they will be used when they are older. They refine their fine motor control, develop their paw-eye coordination, practice how to stealthily approach prey, etc. Play hunting as kittens makes them better hunters as cats, so this behavior is rewarded through the enhanced success of those who are driven to do it. Play hunting is good for kittens. Understanding the many indirect benefits of play has helped biologists understand why mammals are so playful when, on the surface, playing doesn't seem to advance survival or reproduction.

The same reasoning can help us understand the many nonprocreative functions of sex. If an animal gains some benefit from having sex, such as better cardiovascular health, higher sperm count, or enhanced social standing, then the drive to pursue sex would be further enhanced due to the success of that animal in leaving offspring. While the proximate motive of sex is simply the pursuit of the pleasure that comes from satisfying an urge, the ultimate evolutionary purposes of sex are numerous and interesting.

Most of these purposes, as we will see, are social—that is, they have to do with how animals interact with their fellow conspecifics, and this makes sense because sex is an inherently social act (masturbation notwithstanding). For animals that are mostly solitary as adults, we wouldn't expect many nonprocreative purposes for sex. And this is pretty much what we see. Because they are solitary anyway, male orangutans are pretty much only interested in sex with females in estrus. They show little interest in sex with other males or even with females that are not in heat. Their antisocial nature has reduced the experience of sex to procreation only. But most mammals are highly social, and as such, they are surrounded by potential sex partners and opportunities to use sex to advance their goals.

Importantly, the fact that sex has nonprocreative functions means that a purely heterosexual orientation would be an impediment. If many or most of the purposes of sex can be gained through sex with any member of the population, refusing to have sex with half of the population would be a detriment to the animal's success. Staunchly heterosexual animals would lose ground to those who are more broadly attracted. That said, aversion to sex with the opposite sex would bring costs as well by the same logic. For this reason, it seems that sexual attraction on a fluid spectrum of pansexuality is the orientation that would most benefit animals, and this is precisely what we most often see. The more social an animal is, the broader their sexual attraction. The more antisocial an animal is, the more heterosexual they are, since the focus of their sex life is fertilization alone.

As we consider the various nonprocreative functions and benefits of sex, take note of how few of them apply exclusively to male-female dyads.

The Oldest Profession

Any discussion of the nonprocreative uses of sex must include mention of prostitution. In animals, including and especially humans, the strength of the sex drive is a liability that other members of a population can exploit. Most animals will go to great lengths in order to feed their appetite for sex, and there are also other animals that are more "choosy" and can sit back and decide to whom they will grant sexual access and, potentially, at what price. Of course, humans do the same thing. After all, we *are* animals, especially when it comes to sex.

Yes, there are some animals that approach sex as a transaction, and this has a very long history. In many insects, spiders, and even some fish and birds, males must actually "purchase" sexual access through the offering of *nuptial gifts*.[3] Most often, this comes in the form of food, and we call this *courtship feeding*. For example, in nursery web spiders (*Pisauridae*), in order to be granted sexual access, males must approach females with a gift of prey, an insect wrapped in silk. (They even gift-wrap!) While the female is busy eating her gift, the male inserts his mating organ and delivers sperm. There are many variations on courtship feeding among insects and spiders, including examples of animals "cheating" by giving worthless gifts.[4]

In some insects, spiders, amphibians, and cephalopods (squids and octopuses), the males actually *make* the food that they offer the females. Some synthesize the food with their mouths—the salivary glands, to be exact—and some synthesize the offering with their genitals as a package for sperm delivery called a *spermatophore*. And if the spermatophore also "feeds" the female, it is called a *spermatophylax*. These nuptial gifts can be enormous. There are species of butterflies and moths whose nuptial gifts can weigh up to 10 percent of the male's body mass! From the female's point of view, this is a great deal. She gets an enormous meal to sustain her while she prepares her eggs. This is also a very reliable system for ascertaining the quality of the male, considering the burden involved in generating these enormous gifts. But the males benefit from this system as well because, by feeding the female generously, he is working to ensure the success of his offspring.

There are bird species that engage in courtship feeding as well. The great grey shrike (*Lanius excubitor*) is a small bird of prey that hunts rodents, shrews, weasels, frogs, and even other songbirds. A key aspect of their courtship is the provisioning of fresh food by males. Females choose males based in part on the impressiveness of their offering, which, besides making good economic sense, also makes good evolutionary sense. Success in hunting prime calorific foods is one of those proxy indicators of male quality. And, in fact, the quality assessment aspect may actually be the main ultimate mechanism—the evolutionary force—of the nuptial gift. Weighing the quality of the gift accurately measures the quality of the male. Interestingly, male shrikes also procure nuptial gifts for extrapair copulation, emphasizing the quality assessment purposes rather than courtship and pair-bonding functions.[5]

It's not always the males that must pay a price for sexual access. In the Zeus bug (*Phoreticovelia disparate*), the females are twice as large as the males. When it comes time for mating, the male rides on the back of the female for up to a week and she feeds him a waxy secretion while he prepares his sperm for delivery.[6] While it may seem like these males are literally free-riders, the females are actually the choosy sex in these bugs, and, prior to hitching their honeymoon ride, the males fight with each other and physically harass females that haven't yet coupled. Male-male competition, as usual, is brutal, and the waxy meal is part of the prize for which the males are fighting. The females also benefit because, prior to selecting a mate, a female spends a lot of time and energy fighting off the would-be suitors.

Nuptial gifts aren't always food. The spermatophore of the bella moth (*Utetheisa ornatrix*) contains a toxic alkaloid compound used in its defense against predators.[7] This compound is usually found in plants to discourage herbivory and serves much the same function in bella moths, who have evolved defenses that allow them to store it safely in their own cells. The compound is intensely bitter to the taste, making the insects unpalatable to predators.

Interestingly, female bella moths compete with each other to attract as many mates as they can. Successful females are polyandrous, mating with several males, then allowing the different sperm cells to duke it out in their reproductive tracts. But prior to conception, females copulate with the males and receive their spermatophore. The alkaloid compound then covers her body, which protects her from predators and, even more important, is heavily deposited in her egg cells. Because of this, both the eggs and the resulting larvae are chock full of the noxious chemicals.

Polyandry with direct female-female competition is an unusual mating system for insects, but it also makes sense given the unique defensive strategy of bella moths. They protect themselves, and especially their eggs and larvae, through the use of a poisonous chemical. By collecting the spermatophore of many males, a female bella moth can accumulate enough of the toxin to protect her eggs and larvae for the weeks to come when they are most vulnerable. This benefits her evolutionary interests and that of her mates. In addition, the resulting sperm competition acts as a selective force for good genes from the males. A healthy male will generate a large number of robust sperm, while a substandard male will struggle to do so. So what looks on the surface like a payment for reproductive access is actually a highly targeted and clever reproductive strategy in which both sexes participate.

Nuptial gifts and courtship feeding are relatively straightforward to understand in evolutionary terms, and while they do add a transactional flavor to mating, they don't really seem like prostitution since the transaction is directly tied to conception, which definitely isn't the usual goal of sex work. So now it's time to talk about penguins.

Adélie penguins (*Pygoscelis adeliae*) are the most widely distributed penguin species, found around the entire coast of Antarctica and nearby islands. As birds go, Adélie penguins are pretty sexual, engaging in copulation throughout the day, every day, with any willing partner they can find, plus any unwilling ones they can overpower. In fact, the first scientific publication detailing their sexual exploits was deemed too indecent for the Victorian era and was rejected for publication. Instead, the lurid details of sodomy, homosexuality, forced copulation with baby chicks, and even necrophilia were circulated discreetly among the elite community of London naturalists. In fact, so comically prudish was the scientist who first observed them, George Murray Levick, that he actually wrote some of his field notes in Greek so that only "educated gentlemen" could read them, lest they inadvertently spoil the

An Adélie penguin roosting on a nest of rocks

innocence of an unsuspecting reader.[8] Of course, we now know that these penguins are hardly out of step with other animals in the wild.

Despite all the hanky-panky, Adélie penguins form strong pair-bonds, and the dyads tend their nests dutifully and equitably. So they are economically monogamous but sexually promiscuous (like the majority of bird species), and they also live in highly social flocks. They also sometimes pay for sex.

With none of the usual vegetation that birds use to build nests, these penguins use stones. Of course, not all stones are equal when it comes to their suitability as nest components. Too smooth and they will frequently topple, too sharp and they can hurt to sit on, too bulbous and they aren't good for stacking, and so on. This has led to a commodities market for nest stones along the Antarctic coast, where you will notice many unpaired young males hoarding stones that they can trade.

Importantly, these unpaired males are just becoming sexually mature, and not all of them are fertile yet. Given fluctuations in population size and the occasional scattering of stones by strong winds, these stones can be quite coveted. Since they aren't in the nest-building business themselves yet, these young bachelors are pure commodities traders, and what are they generally looking to purchase? Sex. A thriving sex market exists among Adélie penguins with nest stones serving as currency.[9] It has even been reported that, after a particularly satisfying session, a young male will throw in an extra rock. A gratuity, it seems.

Transactional sex has been observed in primates as well. In both Japanese macaques (*Macaca fuscata*) and long-tailed macaques (*Macaca fascicularis*), it has been noted that males will trade grooming for sexual access.[10,11] This example is a little more murky because grooming is an important social and courtship behavior in this species, as it is with most primates. So when a male grooms a female prior to sex, is this more like foreplay? Or is it a transaction? Perhaps words and labels are getting in the way here, as they do throughout the topics of this book. Let's go a little deeper with this one.

In a human romantic encounter, few people would be so cynical as to consider kissing or a sensual massage a payment for sexual access. That's called foreplay, and it's about building trust, arousal, intimacy, and so on. But if we consider events earlier in the evening—fancy dinners, flowers, jewelry or other expensive gifts—these *are* considered by many to be the price that men pay to try to advance the relationship (or just the date) to a place where sex might happen. Is this *paying* for sex? Not exactly, but I am not the first to suggest that men often spend money in the hopes that it will gain them sexual access.

The difference between regular courtship stuff and professional sex work is clear, but that's only because the transactional nature of the sex is explicit. There is a lot of innuendo, gray area, and things left unsaid in between the two extremes. A man buying a woman a drink at a bar may not consciously expect sexual intercourse quid pro quo, but let's be honest and admit that the man is hoping that it is a first step toward reaching that goal at some point in the future. And that possibility is the principal motive behind his actions, even if the man knows that most of these first steps don't get him there. In between courtship and paying for sex lies a spectrum of behaviors, expectations, and, frankly, mind games.

So what about the Japanese macaques? Is grooming just foreplay? Well, foreplay generally consists of physical acts of romantic affection per se, meaning they don't further any ends except romantic ones. That's not true for grooming. Grooming removes parasitic lice and other pests and furthers

Social grooming in Japanese macaques

the health and wellness of the macaques, in addition to the building of social bonds. Furthermore, acts of foreplay are pretty specific to sexual encounters, not other forms of physical affection. There are multiple modes of kissing, for example, and the way we kiss our parents is quite different from the way we kiss our romantic partners (I hope). Macaques groom not only their sexual partners but family members and other social affiliates.[12] So the grooming is not foreplay. It provides an essential service and is not restricted to the romantic context.

But is the sex-for-grooming transactional? When there are many available females, macaque males will spend a lot less time grooming them prior to sex; conversely, when a male has very limited access to females, he must spend much more time on the grooming before a female agrees to have sex.[13] In economic terms, this is akin to the law of supply and demand. Foreplay and courtship behaviors aren't much affected by market forces, but the value of commodities and resources certainly are. With grooming as the currency, when sex is plentiful, it is also cheap. When it is hard to come by, it is expensive. That sure sounds transactional to me.

In common chimpanzees, males have been known to offer meat to females in exchange for sexual access.[14] This too is a little murky, underscoring
the difficulty in sorting out courtship from transactions. For chimps, meat is
not a common food source, but when it is available, it is highly prized. Males
often hunt in cooperative groups, and sharing is expected within the group.
Males will also share with other males in an effort to build allies and advance
one's social rank. But when it comes to females who are sexually receptive,
the exchange of meat is pretty neatly tied to gaining sexual access. That seems
pretty transactional, but there is a complication: males will also share meat
with nonsexually receptive females.[15]

If the purpose of the gift is to win sexual access, why bother giving it to
females who aren't fertile or interested? And since the hierarchy is male-
dominated, female friends aren't really helpful. In some other species, we
might say the males are just being socially generous, but chimps are pretty
shrewd; that kind of generosity would be out of character. So what gives? The
cynical response is that they are trying to get in the good graces of these females in the hopes that it will boost their chances of sexual access later, when
the females *are* sexually receptive. Cynical or not, this is what scientists have
found.[16] Male chimpanzees that share meat with nonreceptive females are
more than twice as likely to be granted sexual access when those same females
are in their next estrus phase. And the effect is specific to the 1:1 relationship,
not an overall trend that generous males get more sexual access more generally.

In my view, the indirect nature of the exchange of meat for sex underscores
the futility in applying human social constructions, such as prostitution, to
animal behavior. But if we look at this the other way around, does the chimpanzee behavior illuminate something about us? Perhaps the chimpanzees are
not "paying for sex" any more than a human being who cultivates a reputation
for being kind and generous and occasionally reaps social rewards for it.
Maybe we're overthinking or oversexualizing this. Maybe within the brutal,
male-dominated chimpanzee society, there are females who quietly push back
and express a preference for more genteel and prosocial males.

We tend to think of animal instincts and behaviors as relatively fixed features, but nothing in nature is static and behaviors can change and evolve.
Remember the mountain gorillas of Virunga from chapter 3? In a single

generation, they totally transformed their whole social structure, and this surely didn't come out of nowhere. There must have been forces at work resisting the dominant regime and poised to push the society in a different direction. The same might be true for the common chimpanzees. Currently, aggressive and insensitive males are winning the day, but there may be subtler forces at work as well, perhaps some leftovers from an outdated social dynamic, or some new raw material for future social evolution. This seems particularly likely given how different the bonobos behave despite being so closely related to the common chimps. Underneath the surface, there may be variety in chimpanzee behavior ready to be selected when conditions change.

There is also at least one case of a monkey directly purchasing sex with cash. Researchers led by Laurie Santos at Yale University have trained a group of captive capuchin monkeys (*Cebus apella*) to use silver coins as currency in a long-running experiment nicknamed Monkeynomics. The coins, used to purchase food and treats, are used in a whole range of experiments to probe how monkeys, and perhaps us, measure value and make decisions in a marketplace. There is much to say about what Professor Santos has learned, and I encourage you to look up her work. Germane to the current discussion, it didn't take long from when the monkeys had learned the currency system for someone to observe a male monkey paying a female monkey for sex. The female, who was not in estrus, turned around and immediately used the coin to purchase a treat.[17] This one is hard to see as anything other than transactional sex. The oldest profession indeed.

When we consider transactional uses of sex, our mind jumps to sex work, but that is only the most obvious and explicit form that it can take. The trading of sexual access for material resources is very common in human cultures throughout history. From a man spending lavishly on a date, hoping to finally get his girlfriend into bed, to an older man funding a younger man or woman to accompany him on vacation as a "travel companion," there are certain unspoken expectations regarding the spending by one member of a romantic pair and the sexual access of the other. Sugar daddies and sugar mommies follow the same understanding.

This is rightly labeled as piggish (or worse) in some contexts, but in others it is purely a matter of a consenting adult pursuing something they want by giving another consenting adult something *they* want. Just like the monkey who wanted sex and was willing to trade a coin for it, and his partner who wanted the coin and was willing to "give it up" in order to get it, no one is harmed when the parties are informed and aware and consent is freely given and may be withdrawn at any point. If you consider that sleazy, that's really your problem.

There are subtler approaches as well. Because the vast majority of psychology researchers are university professors, the sexual market on college campuses has been studied *ad nauseum* and there is an enormous literature on this. Suffice it to say, somewhere around 25 percent of young men, and 15 percent of young women, freely admit that they have attempted to "purchase" sexual access—with no interest in a relationship—using gifts, dinners, favors, tutoring, etc. While unspoken, this marketplace is well understood as such, because the majority of women can recall specific instances when they recognized the niceness of men as an attempt to have sex with them. Shockingly (or perhaps not), 25 percent of women report that they have agreed to sex, even when they recognized it as a quid pro quo for whatever gift or favor they received. At least some of the time, the transactional nature of sex is openly pursued and tolerated, despite the obvious taboo.

As an aside, a small portion of college students—5 percent of men and 9 percent of women—report the opposite transaction as well, in which they attempt to *offer* sex in exchange for something they want. While these percentages are small, they indicate that at least some young people recognize that they can commodify their sexual attractiveness to serve their needs and wants. Again, this may seem icky, but that's on us. Those who are less fettered by cultural taboos around sex don't owe us an explanation.

The point here is that the recognition that sex, both procreative and not, has economic and social value and can be pursued and granted transactionally, and it is in keeping with hundreds of millions of years of animal behavior.

Tricksters

Perhaps it was no surprise that other animals trade resources for sexual access, but you may find it surprising that animals can use sex to trick and deceive. As with the transactional exchanges, we see that the sex drive is something

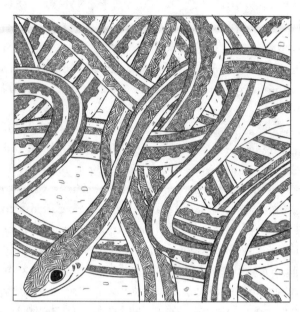

A tangled orgy of garter snakes

that can be exploited and used against us. At the same time, the trickery is still aimed at reproductive goals, either directly or indirectly. Then again, so is all behavior, in a sense.

Garter snakes (*Thamnophis*) are one of the most common snakes in North America, and their reproductive behaviors would be quite shocking to the likes of George Murray Levick, the scientist who first reported on the sex-crazed Adélie penguins. Garter snakes engage in orgies! (I sure hope the herpetologists who discovered that wrote their field notes in Greek!) There is a lot going on in these orgies, so let's take our time.

Like many mammals and reptiles in northern climates, garter snakes hibernate for the winter and have a true "mating season" in the spring, where most copulation and fertilization takes place.* When the snakes come out of hibernation, mating is literally the first order of business. As they emerge from their underground dens, they immediately join the nearest sexual frenzy and form what is called a *mating ball*, a tangled mass of snakes rubbing their genitals together.

---

* Females do copulate throughout the year, however, and will occasionally store sperm that they receive in the fall to use in the spring.

Snakes are *exotherms*, the scientific term for cold-blooded, meaning that they do not generate much of their own body heat, nor do they maintain a steady body temperature. Rather, they get their heat mostly from the environment, and their temperature fluctuates based on the outside temperature and their own rate of activity. Any time you see a reptile basking in the sun, you can be sure that it is trying to warm up so that it can be more active. Cold-blooded animals do generate some of their own body heat from their metabolism, like warm-blooded animals do, but much less so because exotherms have far slower metabolisms than endotherms. This is why a crocodile, while capable of impressive bursts of speed, tires out very quickly.

Scientists believe that their cold-blooded constitution is the key reason that the garter snakes engage in orgies immediately after hibernation. The frenzied mass of snakes creates an insulated reservoir of heat. The heat that is generated from their metabolic activity is shared with the frenzy and insulated within, rather than radiating out to the environment. They are conserving heat while also doing the deed. And who is at the center of these mating balls? The females.

In garter snakes, males outnumber females for reasons we do not fully understand. For every one female, there are five to ten males. The females are also much longer, fatter, and slower, while the males are slim and quick. Males spend a lot of time chasing females, so evolution has rewarded the quick ones, whereas females need to conserve their energy for bearing young. Yes, garter snakes produce eggs, but they do not lay them. They are *viviparous*, meaning they incubate and hatch their eggs internally and then give birth to live baby snakes.

The spring mating balls consist of one or two females surrounded by ten or more males. It's a frenzy all right, but it's also a sausage fest, and the males are furiously competing with each other for access to the female's *cloaca*, her genital opening. The fastest and best wrestlers earn the most paternity. The competition generates one of those male-based quality-control systems, and the females can accept sperm from a large number of males, diversifying the genetics of her clutch.

Males emerge from hibernation before the females, and they use this time to warm up their bodies, from both internal and external sources, so that they are ready to swarm the females when they emerge. The males recognize the female when she emerges because of a female-specific pheromone she emits, like perfume, if you will. As the horny males emerge and largely ignore each other, they are poised to pounce on the first perfume-laden snake that they find, hoping to be at the center of the mating ball and have the best chance to sire many offspring.

And this is where the deception comes in: scientists have found that some males are capable of emitting the female pheromone.[18] When they emerge from hibernation smelling of perfume, nearby males are fooled and swarm the female-smelling male, attempting to mate with him. Obviously, a male cannot be impregnated by another male, so what is to be gained by this behavior?

Are these just cross-dressing gay snakes, luring unsuspecting males into bed with them for fun? Maybe. But the more sensible explanation is that these males are using sex to trick the other males and gain a competitive edge over them. If a perfume-wearing male snake encounters other snakes when he emerges, that automatically means that he's late to the party and needs to catch up. Snakes emerging from hibernation are cold and sluggish, making them terrible competitors in a mating ball as well as vulnerable to predators. (Slow springtime garter snakes are a favorite meal of crows, for example.) So by wearing perfume, these males trick the other males into swarming them, thereby protecting them from predators and warming them up quickly. Within a few minutes, the jig is up, but by then the perfumed males are ready for the fiercely competitive orgy with a true female. They literally *steal* body heat from them.

Second, the fooled males use up valuable energy and must then recharge because exotherms cannot sustain strenuous activity for very long. The female-mimicking snakes tire the others out, thereby giving them a competitive advantage for a little while.

And it works. Scientists have found that the males who emit female pheromones sire more offspring when they mate.[19] Once they're warmed up and ready, they ditch the perfume and go after the females. This is a successful alternate mating strategy. You may remember that I expressed my skepticism of other examples of so-called female mimicry in chapter 3. But in the case of the garter snakes, deceptive cross-dressing is the simplest explanation that is consistent with the data. The key distinction, at least for me, is that the males respond differently to the same snake based on whether or not he is producing the perfume. When he is, they mob him and have sex with him; when he's not, they don't. By mimicking females, he both tires out and steals body heat from his rivals, and once he does that, he uses this advantage, removes the drag, and pursues the females.

Given how transient the cross-dressing is, I would not consider this a distinct gender, but it is clearly another example of the creative ways that animals can play around with sex and gender in order to stick out in a crowded field. Scientists have even worked out the molecular details of this cross-dressing, and it is mediated by a hormone you've probably heard of: estrogen.[20] And the fact that only a minority of snakes pursue this strategy is part of what makes it successful. If all the snakes did it, the whole mating system falls apart. But if only some snakes do, they can get away with it. It's important to note that being a strong and fast male is still a good competitive strategy. Those males do well in the mating balls no matter what. So what the garter snake population is left with is a balance, an equilibrium between the diverse types of gendered mating strategies. By now this should sound familiar.

Speaking of animals that do something gay in order to gain advantages over other males, let's discuss the flour beetle (*Tribolium castaneun*). Yes, these are the insects that frequently infest flour and other dry groceries and whose larvae look like mealworms. Like the bed bugs we met in chapter 3, these insects have been known to engage in male-male mountings, but, unlike the bed bugs, copulation in this species is not traumatic. Males and females are quite dimorphic, so it's doubtful that the males are simply confused. They mount other males on purpose, and the reason why is pretty tricky indeed.

Like most other arthropods, flour beetles don't have external genitalia. Like most birds and reptiles, both sexes have a reproductive tract, and males inseminate females by lining up their genital openings and squirting their sperm inside the females' genital tract. Their mating strategy is promiscuous. The females mate with many males, leading to sperm competition in her reproductive tract. Females benefit from mating with many males by diversifying their genetic stock and allowing the resulting male-male competition to select for good genes. For males, the best strategy is also to mate with as many females as they can, pushing other males out of the way whenever possible.

Some male flour beetles take things even further: they mount and deliver sperm to other males. This would seem counterintuitive because it wastes sperm where it can't inseminate and wastes time and energy while other males could be reaching nearby females. It turns out that neither the sperm nor the energy are wasted. When a male gets another male's sperm deposited in his reproductive tract, it can survive there for several days. When he encounters a female and mounts her, he delivers not only his own sperm but also that of the male that previously mounted him.[21] So by mounting several other males, a male can turn them into his unwitting sperm-delivery foot soldiers, allowing him to inseminate females *that he doesn't even personally encounter*.

This is yet another case in which same-sex sexuality boosts biological fitness, but it's also an example of how sex can be used in deceptive and combative ways. This particular strategy of using the reproductive tract of other males as a vehicle for delivering one's own sperm wouldn't be available to mammals since we, you know, have penises that can't usually be penetrated by other penises (although I am sure that Rule 34 applies here). However, in nearly all birds, reptiles, and amphibians, and the majority of fish as well, males and females have very similar-looking genital openings. In fish, it is called a *genital pore*. In birds and reptiles, it is called a *cloaca*, which is a single common opening for the urinary, digestive, and reproductive systems.* I suspect that the indirect sperm transfer through male-male mounting that we see in flour beetles is found in some of those animals as well. This hasn't been documented, but it also hasn't been looked for yet. Scientists only find what they look for.

Lovers, Not Fighters

After covering the transactional and deceptive uses of sex, we need a good palate cleanser to remind us that animals aren't all cheats and swindlers. Some of them are lovers, not fighters. Sex is used by many animals in just the same way it is used by us: to draw two or more individuals closer together. As we explore this function of sex, we are limited to birds and mammals because only in these groups do we observe the neurohormone-based attachment system that is activated by sexual activity (that we know of).

---

* *Cloaca* means "sewer" in Spanish. You're welcome.

Importantly, the use of sex to strengthen social attachments is not limited to pair-bonds and is even active in species that do not form pair-bonds at all. Take, for example, the bonobos. Although they do not form 1:1 reproductive pair-bonds, they do form very close social attachments to their friends and allies, and among family members (with the exception of mothers and adult sons), these attachments are strengthened by sex. Some biologists tend to think of the bonobo social structure as kind of a fluke, with all its rampant sexuality, but, fluke or not, they are our closest relatives, and 98.5 percent of our coding DNA sequences are identical. We cannot write them off so easily.

Currently, in human societies, this particular use of sex is mostly restricted to those in romantic relationships, whether or not they are long-term, sexually exclusive, etc. For example, the majority of married couples in open relationships have rules around extramarital sex designed to prevent or mitigate the attachments that can form between sexual partners. (This is different from polyamory, where such attachments are allowed.) The fact that swingers and those in other forms of "open marriages" recognize the tendency of regular sex partners to form strong emotional attachments underscores the fact that humans have an intuitive understanding that sex often leads to emotional intimacy, especially when it is repeated. Because we covered social monogamy and pair-bonding in the previous chapter, we can now focus on how sex can be used to forge social bonds within larger communities of animals.

In the last chapter, we discussed the most abundant subspecies of lion, the Tsavo lions, which exclusively form prides in the one-male, several-females harem lifestyle. But in other regions of Africa, there are prides led by multiple males. These multi-male partnerships are called *coalitions* and form through a combination of both competition and cooperation. Within the larger population of lions, individual prides go through periods of stability punctuated by fission-fusion events. Male lions periodically challenge each other and fight, but they also forge teams that cooperatively work together to dominate a pride and fight off other groups of males.[22] Members of a coalition work together, but there is also tension because paternity privileges follow the dominance ranking. Young ankle biters are anxious to move up the ladder, and aging alphas cling to their position as long as they can.

During times of stability, however, the males of a coalition live in relative harmony and build trust and attachment with each other.[23] And how do

Two male lions in affectionate contact

they build that trust? You guessed it: sex. Male lions within a coalition are constantly having sex with each other, even more often than they do with the females, even though paternity rights are coveted by all. The likely explanation is that the females are much less receptive to sex than are the males. Since the males are almost always "in the mood," they have a ready outlet for their arousal: each other. As usual, this was missed by scientists for decades because they interpreted the behavior as either dominance challenges or play wrestling. Indeed, some of that may also be going on, but considering that the "wrestling" includes anal penetration and orgasm, it's safe to say that they are having sex. Males in a coalition are tightly bonded social partners and display all the signs of affection and mutual care that Tsavo lions (and silverback gorillas) show the females in their harems.

The coalition system is a real win for most male lions because when only one male is tolerated at a time, the majority of males are unsuccessful.[24] For the alpha, coalitions are more of a mixed bag, since he has to share paternity with his subordinates or else they will turn on him. Somewhat making up for this, an alpha is able to maintain his dominance for longer with a coalition than he would on his own, generally speaking. The question of which system is more successful overall depends on perspective. What is good for the alpha and for the other males isn't always the same thing. The fact that different lion communities navigate this tension differently speaks to the inherent flexibility of the sex lives of lions. Do the coalitions of male lions bonded by sex have any relevance to the human experience?

In ancient Greece, homosexuality was not at all taboo, and the power of sex to forge close-knit bonds in the military setting was widely recognized.[25] The strong bonds strengthened by sex between military partners is discussed in a variety of ancient texts, including writings by Socrates and Aristotle, and later by Homer in Rome. In *Symposium*, Plato spoke highly of the practice, writing that "no man is such a craven, that the influence of love cannot inspire him with a courage that makes him equal to the bravest born." This was not without controversy. The military commander Xenophon of Sparta was particularly critical of the practice, but he didn't disapprove of the homosexuality itself. Rather, he found it crude to exploit the beauty and power of sex for such a utilitarian military purpose. He also disapproved of the formation of military ranks based on anything other than talent.

The Sacred Band of Thebes was an elite force of Greek soldiers formed entirely of pairs of male lovers and commanded by the general Epaminondas, who himself had two lovers among the ranks. This strike force is credited with turning the tide against Spartan domination of Greece in the fourth century BCE before eventually itself being defeated by the Macedonian king Philip II. Epaminondas and one of his lovers were killed in the battle and buried together with full honors. Following the defeat of the Sacred Band, the practice began to decline, but its value was recognized and later encouraged by the son of Philip II, Alexander the Great. Alexander, of course, had his own lovers, both male and female, and recognized how his soldiers fought with greater passion and focus when their lovers' lives were also on the line.

Male-male bonding through social interactions, including sex, have been observed in some animals as a successful cooperative reproductive strategy. In many mammals—perhaps most—the males are rivals, not friends. Among walruses, deer, and many others, males are in constant competition and see each other purely as threats. But in highly social species, including many primates, close affiliation and cooperation has evolved as a strategy in which all or most males are actually better off than if they were fighting each other for dominance and exclusive mating access. The strengths of different social strategies are hard to measure and compare in the wild because, generally, each species behaves in a certain way; and comparison to other species isn't simple because many other variables will be different too, not just the way the males interact with each other. Nevertheless, in 2014, scientists discovered key insights by studying male-male social bonding in Guinea baboons.

Guinea baboons are an excellent animal for answering questions about bonding because their social structure is mixed with different kinds of smaller groups nested within a larger community. Within this are one-male harem-style groups, multi-male/multi-female groups, and lone "bachelor" males, usually juveniles that haven't yet found their place. Unsurprisingly, the males living in mixed-sex groups show less aggression toward each other and more displays of social affiliation, including grooming, food sharing, and sex. The surprising result that the researchers found, however, is that, as a male baboon grows older, stronger, and more dominant, he switches his social affiliations toward females and does his best to win a harem of his own.[26] Although only a small portion of males will reach that goal, the males leading a harem enjoy the most reproductive success.

So for a male Guinea baboon, life is a game of climbing the social ladder in the hopes of reaching the top position—the sole male leader of a harem. However, the ladder that must be climbed is one of affiliation and cooperation with other males, including fighting against other male coalitions on occasion. The bachelor males that are not integrated with other males suffer a much higher rate of mortality. On the other hand, the males that associate with other males but are of low rank obtain less reproductive access to females. Female choice is the key determinant of male reproductive success in baboons, and females tend to choose the high-ranking males that are popular and respected by other males.

The best strategy is to be friendly, gregarious, and generous with other males, which earns the respect of both males and females. The key to the friendly affiliation of males is—you guessed it again—the pacifying effects of sex. Sex creates a form of social glue that bonds males together, reduces aggression, and promotes prosocial behaviors such as food sharing and grooming. Males that perform well in this social environment earn the most paternity.

The other thing worth noting about the Guinea baboon lifestyle is its diversity.[27] Although there is a clear path for the *most* reproductive success, there are a variety of ways to live and thrive. There are lone males on both ends of the social spectrum, there are friendly males as well as more dominant/aggressive ones, and females are the social directors of the whole show, also displaying a mix of cooperation and affiliation as well as competition in a dominance hierarchy, particularly within a harem. This sets up a diverse social environment that is poised to evolve in different directions if and when conditions change.

Choking the Chicken

I'm not sure why, but most people are genuinely astonished to discover that animals masturbate. But this shouldn't be a surprise because, finally, we as a culture admit openly that nearly all people do it, and anyone with a dog has surely noticed that they do it—so why are we surprised to learn that most animals do too? Personally, I think the reason that people are confused by animal masturbation is that they don't think of animals as doing things that are just for fun and without any "larger" purposes. This misses the point in two important ways.

First, animals absolutely engage in behaviors that are, at least on the surface, purely for fun. Why? Because behaviors are often much more than they seem on the surface. Yes, animals play because they enjoy having fun. But play behaviors also serve several other functions and are good for the animals in a variety of ways. There are social benefits, psychological and neurological benefits, cardiovascular benefits, hormonal benefits (stress reduction), benefits to motor control and coordination, and many more. We play because it's fun, but evolution has linked play to pleasure because it's good for us in very clear and specific ways.

And second, just as with play, masturbation does at least sometimes serve other purposes directly. Animals need not be aware of these additional purposes and benefits, and in fact they never are (that we can tell). All they know is that it feels good. So for the purposes of our discussion, I want to first entertain you with some facts and stories about animal masturbation, and then we can explore the possible benefits that masturbation brings and consider if any of those purposes apply to humans as well.

YouTube is a veritable treasure trove—or cabinet of horrors—of animal masturbation videos. With just a couple of search terms you can find a video of a raccoon touching himself, countless monkeys doing the deed, and dogs humping just about every object you can think of. You can also find animals stimulating themselves orally, as well as using various implements. There is a horse stimulating herself with a fence pole, a chimpanzee rubbing a frog on his penis, a river dolphin penetrating a dead fish, and even a young tortoise humping a croc (the footwear, not the animal). These are all images you now have in your head. You're welcome.

Masturbation in animals has also been the subject of scientific study.[28] Some of the species in which masturbation has been carefully observed, described, and speculated about include horses, cats, donkeys, lions, kangaroos, cows, walruses, bears, and even porcupines (ouch!). Caribou and antelope achieve sexual gratification by running their antlers against hard objects. Orangutans have even been known to fashion objects—or should I say toys—to aid in masturbation, such as fleshy fruits, while macaques have been known to use makeshift phalluses.[29] Some animals, such as dogs and horses, rarely orgasm while masturbating, while others, such as chimpanzees and cows, almost always do.

So why do all these mammals masturbate? Well, why not? Not all behaviors have to be tied to some grand evolutionary purpose. Why are so many humans unable to stop biting their fingernails? Who knows? There may not be complete or satisfying explanations for all the behaviors we find perplexing. The answer could be as simple as animals masturbate because it feels good and it is an unavoidable consequence of linking up sexual behaviors to the pleasure centers of the brain. Masturbation may be an inevitable and harmless quirk of animal sexuality.

Nonetheless, some benefits of masturbation have been proposed. For males in species that engage in sperm competition, there is a clear benefit: fresh semen seems to work better.[30] Therefore, "cleaning out the pipes" may actually boost the quality of both sperm cells and the semen that activates them. Sperm competition occurs in species in which the females mate with many males. Within the female reproductive tract, sperm have a very long way to swim relative to their body size. It's the equivalent of a human running about thirty kilometers (nineteen miles) nonstop at high speed, but their actual path is longer than that because they don't exactly swim in a straight line directly to the egg. Sperm cells meander aimlessly and swim in a corkscrew fashion of ever-expanding right-hand circles.* Very few sperm end up anywhere near the target.

With such a long way to go (and such poor aim), the key to successful fertilization is to ejaculate *a lot* of sperm. In humans, the average ejaculation releases *hundreds of millions* of sperm cells. That's a lot, but it's nothing compared to some other species, including our closest relatives, both species of chimpanzees. Because female chimps mate with so many males, sperm competition is intense in these apes. That's why chimpanzee testicles are enormous and their ejaculates are voluminous. Masturbation helps also because it clears out old and defective sperm cells, as well as stale seminal fluid. This allows the next ejaculate to be as fresh and speedy as possible. Indeed, because sperm cells are stored with inhibitors to keep them inactive, the longer they are in storage, the more sluggish they are when they are released. It has been shown in humans and rhesus macaques, for example, that regular masturbation boosts sperm quality.[31]

For females, masturbation can be *essential* to reproductive health.[32] Some mammals employ what is called *induced ovulation*, meaning that eggs are not released on a regular schedule, as with the human menstrual cycle, but rather, eggs are only expelled from the ovary when a female is stimulated by sexual activity. This is especially useful for species that live in low population densities. If an ovulation comes and goes with no male around, it is wasted. On the other hand, mature eggs can hang out in the ovary for a little while, but if they are not eventually ovulated, they can develop into a painful cyst. Domestic cats are induced ovulators, and anyone with an intact female cat has probably observed the painful experience that occurs if she doesn't eventually have sex when she is in heat. If there are no males around to help her out, what can be done? (Well, you can purchase a small device to help if you want. It looks like a Q-Tip. Have fun. Otherwise, the humane thing is to have your cat spayed.)

---

* When the French translation of *Human Errors* arrived at my house, I was quite surprised that they titled the book *Sperm Only Turn Right,* inspired by this fact.

It's no surprise that many cats and other induced ovulators have figured out that masturbation relieves the abdominal pain. Some of the other species that ovulate this way are ferrets, rabbits, camels, llamas, bears, and wolverines. In some of these species, such as black bears, manual stimulation is not enough. A male bear must actually be physically present and involved in order for a female to ovulate. Koalas have it even worse: they need a male, genital stimulation, and ejaculated semen in their vaginal canal in order to ovulate. For most other induced ovulators, masturbation is enough. It is also worth noting that even in species in which stimulation is not *required* for ovulation, it can help bring it on. In humans, for example, in the days around ovulation, a well-timed orgasm actually helps accelerate the hormonal changes that lead to ovulation. So now we have a benefit of masturbation in males (sperm quality) and in females (induced ovulation).

Another proposed function for masturbation in both male and female animals is to practice mounting behaviors for later in life, similar to how puppies play-wrestle as practice for dominance challenges they will participate in as adults. Remember that because sex is not just for procreation, practicing sexual behavior (in order to perform well later) could benefit an animal in all the various ways discussed throughout this book. By practicing the mounting and thrusting of sex, young animals develop the necessary muscles and refine their motor control and coordination. There is not any direct evidence that this actually helps, because it's very hard to design an experimental way to test this hypothesis.* But there is another function of masturbation that we *can* measure.

In 2010, Professor Jane Waterman published an exhaustive look at masturbation in African ground squirrels.[33] Before we get to the reason why this species jerks off so much, I think it's worth a paragraph or two to appreciate their masturbatory habits because they dispel a few myths that still shroud this topic. In the twenty male squirrels that Waterman examined, *all of them* masturbated regularly. Interestingly, the males masturbated *more* when the

---

* Even in nonsexual contexts, it is very difficult to scientifically test the validity of the play-as-practice theory because depriving an animal of play experiences (to observe if they are less proficient at a certain behavior later in life) is impossible to do without also depriving them of other things known to be important, such as social interaction and physical activity. For this reason, animal behaviorists are tentative in their support for play-as-practice.

females were in estrus. This is also when the squirrels have the most sex. So during the mating season, when sex is the most available, is also when the males masturbate the most. In addition, the squirrels tended most often to masturbate *after* sex. This is also surprising because people so often think of masturbation as something that males do when they *can't* get sex. And last, Waterman found that the more dominant males masturbated *more often*, once again bucking the common human expectation.

Far from something that frustrated males do when they can't score a sexual partner, masturbation in ground squirrels is an accompaniment to sex. The more sex a squirrel has, the more he masturbates. This led Waterman to her hypothesis that, at least in the males of this species, masturbation is a form of hygiene that is used to clean one's genital anatomy to prevent the spread of sexually transmitted infections (STIs). Oh, and one more piece of evidence for this: in all 105 masturbation sessions that Professor Waterman documented, the squirrels used their mouths for gratification and consumed their ejaculate. With its many antibacterial components, saliva makes a good cleaning solvent. And second, these squirrels are desert-adapted, so the consumption of masturbation-induced ejaculate would be consistent with those adaptations, so as to prevent unnecessary water loss.

African ground squirrels have a mating season, and a typical female can mate with ten males in a single afternoon sex romp. A nasty STI would spread very quickly through the population before their immune systems could stop it. While their short lifespan means STIs are unlikely to be fatal, they could be devastating for fertility. By cleaning the genitals inside and out with some antibacterial saliva, these squirrels may be protecting themselves, their partners, and even the whole community. This function of masturbation would explain why it tends to happen after sex, is more common during the mating season, and is most frequent among the males who have the most sex. Biologists now propose a similar function for masturbation in several other species.

Each of the various purposes and benefits of masturbation may or may not apply to us, but that doesn't really matter. Each species may derive its own benefits or none at all. And I should admit that not *all* mammal species have been observed to masturbate, and some have only been caught rarely. It could be that, if none of these benefits exist, and the species is not that sexual in the first place, masturbation will be rare. What is important, however, is that research on humans has made it abundantly clear that masturbation is not just normal but beneficial to psychosexual development. As our bodies first mature, masturbation is how we familiarize ourselves with the new functions our adult bodies have. It's how we discover what we enjoy, both for arousal and physical touch. It's an important part of our sexual exploration and something that we should talk about more openly.

Masturbation has also been found to be beneficial for mental health, stress reduction, and even cardiovascular health and physical well-being. Masturbation has even been shown to boost the immune system, likely through the endorphins and other neurochemicals that are released during sexual activity. Self-pleasure can help us relax, help us sleep, help us focus, and boost our sexual interest and libido. For men plagued by premature ejaculation, masturbation can help them regain control; and for men experiencing erectile dysfunction, solo sex is almost always an important part of the recovery. For women, masturbation can help with menstrual cramps. For both men and women, masturbation can boost sexual performance and enjoyment, as well as body image and sexual self-esteem. Masturbation is good for us in all kinds of ways, so it's no surprise that we have evolved to enjoy it.

Everything Is About Sex, Except Sex*

There is a saying, often attributed to Oscar Wilde: "Everything in the world is about sex, except sex. Sex is about power." Wilde almost certainly didn't say this because the use of the word "sex" to mean "sexual intercourse" didn't begin until decades after his death. Nevertheless, this is a profound insight, and most psychologists will tell you that there is a complex truth behind this assertion.

Much of what animals do revolves around sex. It occupies a great deal of their time if you consider the pains they take to get it, the time they spend doing it, the efforts they engage to make it successful, and so on. Consider the sockeye salmon that will swim for hundreds of miles (upstream!) in order to spawn once and then die. Fish don't even get to enjoy interactive sex like we

---

* This section has been lightly adapted from chapter 5 of *Not So Different: Finding Human Nature in Animals* and appears with permission from Columbia University Press.

do—they just squirt their sperm and eggs, and that's it. And yet, look at the lengths they go to! Look at the elaborate coloring of the golden pheasant—all of that flash and flare in order to get a mate.

For mammals, sex is much more fun, and, understandably, we spend even more time doing it and trying to do it. Although this chapter is dedicated to exploring the many functions for sexual activity that do not directly lead to conception right then and there, it is important to note that these behaviors are still directed toward reproduction indirectly, because *all* animal behavior is directed toward the advancement of reproductive goals. Reproduction is the ultimate goal of all animal life.

Even sexual activity that doesn't seem connected to reproduction might actually be, at least in part. To wit, sex is used to establish relationships and alliances and secure one's place in the power structure of the group. Sex is used to exert and establish dominance. Sex is used to play tricks on, distract, and exhaust competitors. Sex is used to resolve disputes and maintain group cohesion. Sex is used to strengthen pair-bonds and promote good family life for raising offspring. All of that does have impacts on an animal's chances of successful reproduction, if indirectly.

If we are talking about "sex" as everything connected to mating, repro- duction, and child-rearing, then yes, just about everything an animal does is about sex. With this extended definition of "sex," it turns out that the "pur- ists" might be more correct than we originally let on. We began this chapter by saying that sex is about a lot more than just making babies. What I meant was that sex is about a lot more than *conceiving* babies. If we consider pair- bonding, herd cohesion and harmony, social hierarchy, or just plain having fun, the benefits and purpose of sex are built around providing for the future of the species. It's all intertwined and cannot easily be separated. When it comes to animals in the wild, it *is* all about sex.

The question is: Are humans any different? Remember that I am not just talking about conception, and I'm not even just talking about the sex act itself. For many of us, much of our lives could be summarized as making ourselves attractive to a potential mate (physically, financially, socially, and so on), searching for and selecting that mate, establishing the resources and structure to build a family home, having children, caring for those children, preparing them for their own success in this cycle, retiring from our own family building and putting resources into the success of our children's children, and so on. These need not be our conscious motivations. Drives and urges often work subconsciously.

I'm also not saying that we don't do other things along the way, but the major events and aspects of our lives involve our attempts to contribute to the next generation of our species the best that we can. Some of us are better at it than others. Others opt out of this altogether. Still, we all live in this herd together. It takes a village to raise a child, and, like it or not, we're all part of the village.

Chapter 6
Family Values

For many, the current trend in young people today eschewing monogamous heterosexual dyads in favor of more poorly defined and less restrictive relationships seems like a departure from the natural state of romantic relationships for humans. Of course, as I see it, this is actually a *return* to a more natural state of affairs. (Pun intended.) To make that argument, I reached back a billion years to the very origin of sexual reproduction and carried the discussion through the evolution of animals, sociality, and child-rearing, to emphasize that the enduring theme of sex, gender, and sexuality—indeed the theme of life itself—is creative diversity. Humans are a product of that billion-year legacy, so why should we expect anything other than wondrous diversity in how we engage sex and gender?

That said, there is simply no denying the primacy of heterosexual monogamy in most cultures around the world. Is this telling us that, somehow over the last few million years, humans have evolved into monogamy and strict heterosexuality as our natural biological state? Indeed, this is precisely what social conservatives would have us believe: that only heterosexuality is natural for humans, and that permanent monogamous marriage is not only the proper way to make a family but the bedrock institution around which society is built. For many of the people who feel that way, our exploration of sex and gender in other animals is irrelevant because they don't believe in evolution anyway. Humans were created differently and that's that. These supposed "laws of nature" were then passed down to us as though they were fixed universal constants, like the speed of light in a vacuum or the Ideal Gas Law.

But those of us who accept the overwhelming evidence of universal ancestry are left to wonder: How did humans end up with such a limited experience of sex and gender compared to other animals, including and especially our closest relatives? After all, each one of us is the biological offspring of two people who had sex, who were themselves the product of two people who had sex, and so on back through the countless eons of sexual creatures doing what sexual creatures do. If monogamy and strict heterosexuality aren't biological imperatives, how did they become so dominant?

To answer *how*, we must first consider *when*, and then we will get clues about *why*. As we will soon see, the contraction of human sexual relationships was a much more recent development than most people think.[1] This switch only occurred as human history marched toward centralized societies, which consolidated power and resources and collapsed cultural diversity. Even in those parts of the world where nation-state civilization has its deepest roots (China, the Middle East, Egypt, Mediterranean Europe, and the Indus Valley), we are talking about just a few thousand years, which is far too recent to be the product of biological evolution. In the rest of the world, the history of centralized governmental bureaucracy is even more shallow. Prior to the rise of colonizing empires and authoritarian governments, and the cultural steamroller that always accompanies them, a different way of life flourished. Or, I should say, different *ways* of life. In keeping with the grand traditions of our fellow mammals, diverse human groups have explored myriad ways to survive and thrive as sexual creatures with strong social bonds. It turns out human constructions about sex and marriage are not written in stone—or in DNA—and could very well have been constructed very differently by human culture.

And indeed they have been.[2] As universal as it may seem now, the stricture that marriage is a one-man, one-woman arrangement that has dominated much of the world for the last few centuries is a relatively recent social innovation that spread mostly by force, to be perfectly frank. Prior to the dominance of monogamous heterosexual marriage, other kinds of relationships—almost any kind of arrangement you can think of—existed in various cultures around the world, including in Europe, the cultural heart of what is now "the West."

This is not to say that the heterosexual dyad is itself an aberration or a recent innovation. Of course not. Straight couples have been straight coupling since time immemorial. What *is* new and quirky is the idea that this is the *only* type of romantic or family arrangement for which humans are suited. During the marriage wars of the 1990s and early 2000s, the favorite refrain of social conservatives was that marriage was between one man and one woman. The appropriate response to this assertion is, *Yeah, but only because we have said it* must *be so.*

It is true that the majority of the population are cisgender, and the majority of men are sexually attracted to women and vice versa. It is even true that this sexual attraction is largely driven by inherent biology. However, to go from there to permanent monogamous marriage is quite a leap indeed. And it is another leap entirely to maintain that this is the *only* route any of us can take. Neither of those leaps are supported by the biological evidence. They are purely the product of *cultural* evolution, not biological evolution.

Of course, cultural forces are not necessarily bad and there is nothing wrong with marriage per se. We construct lots of things that are not encoded by our genes but that are worth holding on to. Money, laws, and literature come to mind. But since the topics of sex and marriage are fraught with conflict, their natural history is worth scrutinizing. And scrutinizing is what we are here to do. As we learned in chapter 4, in animals, social or economic monogamy is an entirely separate matter than sexual monogamy, which is exceedingly rare. But humans have attempted to join these two into a single concept called "marriage" and then declare it permanent and sacred. But if we strip off the cultural accoutrement and focus simply on the relationship itself, we find that humans are not so different from our animal cousins in terms of how we pursue sexual partners and build family relationships.

As we explore human cultures through different times and places, we find many permutations of the monogamous relationship, as well as cultures that lack almost any trace of what we would recognize as monogamy or even dyads. In the world today, variations on traditional monogamy are very much on the rise, as is the abandonment of dyads altogether. Contrary to the common reactionary response, these variations are *not* new or exceptional. They are in keeping with a long history of relational experimentation, flexibility, and diversity among our ancestors. Let's take a look.

Foraging for Love

Given our common ancestry with the other African apes, and other primates and mammals more generally, it is beyond argument that most of our long line of deep ancestors were not monogamous heterosexuals. That is simply not a common way for primates to live. However, for most humans alive today, our recent ancestors, the ones we can name or imagine, *were* monogamous heterosexuals, or at least they were expected to be. So when did this shift occur? Does monogamy date back to the origin and evolution of our species, hundreds of thousands of years ago in the grasslands of Africa? Or is it a more recent phenomenon? This question is easier to answer than you might think because, while modern culture is dominated by globalized civilization, pockets of a more ancient way of life still persist in various places around the world. And, almost without exception, none of them engage in the kind of restrictive sexuality that accompanies modernity and industrialization.

While recorded history covers only about 5,000 years, most of it incredibly incomplete, our species has existed in more or less its present biological form for about 200,000 years. For nearly all that time, humans existed in small bands numbering less than 250 individuals per group. This lifestyle is sometimes referred to as *foraging* or *hunting and gathering*, and it is highly nomadic. Humans did not begin to build permanent settlements until farming was invented, an event that occurred nearly simultaneously in at least seven different locations around the world 10,000 to15,000 years ago. It was agriculture—the cultivation of crops and livestock—that led to the seismic shift in how our species lives, transforming the small roving bands of foragers into bustling villages and towns, swollen with people. However, this shift was so recent that its effect on our biological nature—the evolution of our genes and brains—has been almost nil. We have the brains and bodies of the foraging hunter-gatherers that comprise 95 percent of our human ancestors.

As a quick aside, the life of foraging peoples, and thus the lives that our prehistoric ancestors lived, has been greatly misunderstood and misrepresented.[3] In our usual chauvinistic way, we had presumed that precivilized life was "nasty, brutish, and short," and the people were usually starving, sickly, and terrified.* All evidence points to the exact opposite. Prior to the invention of agriculture, the accumulation of material resources was simply not a major pursuit, and people were tied to neither land nor things.[4] Foraging was for subsistence, not profit, and daily life was far less demanding than the forty-hour-per-week slog that we now know.

Small groups meant that infectious diseases were a rare annoyance. Diets were more closely optimized for our body's needs (or, rather, our biology had nicely evolved to the foraging diet), which means that cardiovascular and metabolic conditions, such as obesity, heart disease, and type II diabetes, were virtually unheard of. Life was much more relaxed and leisurely, vibrant and socially interactive, and more time was spent on hobbies and gossip than on the hard work of survival. Oh, and hunter-gatherers also had (and still have) a lot of sex.[5]

Because of the unyielding spread of the agricultural lifestyle and the tendency of civilizations to irreversibly assimilate cultures as they expand, the foraging lifestyle—in a sense, the natural state of our species—has been nearly eradicated from the planet. But not completely. There are around thirty full-fledged societies, spread across every inhabited continent, that continue to live in their own unique foraging lifestyle. Although they live a traditional life, these societies have been influenced by the industrial world around them. To various degrees, they have incorporated modern tools, clothing, and subsistence techniques. But make no mistake, there are still tens of millions of human beings living some form of the foraging lifestyle.

On the other hand, there are somewhere between a hundred and two hundred tiny tribes of so-called "uncontacted peoples" comprising around ten thousand individuals in total. These tribes live a *purely* foraging lifestyle entirely unperturbed by modernity. Most of these uncontacted tribes are in the Amazon basin, but a few others are found in other parts of South America, the impenetrable mega island of New Guinea, one in Indonesia, and two

---

* This view of precivilized human life was most famously described in Thomas Hobbes's magnum opus, *Leviathan*, which argued that government and civilized society were necessary to restrain the worst instincts of human nature.

in the Andaman Island chain off the coast of India. These people have had very limited contact with outsiders and live as close to a "state of nature" as any other humans left on the planet.

Among the thirty large forager societies and the hundred or so uncontacted tribes, you will find many different kinds of marriage and family arrangements—called *mating systems* in the anthropology literature—but almost none of them match the currently dominant model of permanent, sexually exclusive, heterosexual marriage. As we will see, there are almost as many different mating systems as there are societies, leading us to the conclusion that every society does sex, marriage, and family life in their own way. Although there are some common themes, sexual monogamy is not among them. A quick tour through a few of them will illustrate.

The Nukak people are a tiny tribe of foragers indigenous to the Amazon rainforest in central Colombia.[6] They are highly nomadic and rarely stay in the same location for more than three nights in a row. Their mating system does involve what we recognize as a marriage, with dating, courtship, and even the requirement of mutual acceptance by the betrothed families. However, marriage is polygynous because men can take two wives if they wish, and occasionally three or four. But there is another twist: when a married woman is pregnant (and marriage always precedes pregnancy), she is temporarily allowed—encouraged, in fact—to have sex with other men, but only the relatives of her husband.

This temporary polyandry is practiced due to a belief that it aids the growth and health of the developing baby. The more seed that is planted, the healthier and stronger the baby will be. Accordingly, the paternity of the child is considered to be shared equally among all the men that contributed. In Nukak society, every child has many fathers. This system of distributed paternity forges close-knit bonds among the family of the father's child. It also ensures that all children have multiple men invested in their thriving. Combining the temporary polyandry with the marital polygyny results in a complex web of family relationships. This promotes a broad sense of belonging and social cohesion. Although the parents are at the helm, the village raises the child.

In the nearby region of Venezuela and Brazil, one of the largest Indigenous forager groups is the Yanomami, a loose collection of more than thirty-five thousand individuals spread among some two hundred and fifty villages.[7] While not an uncontacted tribe, the Yanomami live in a traditional foraging lifestyle that has been only minimally influenced by outsiders. Each village operates as an extended family, with cooperative parenting and mutual aid and support. Family life is marked by a mixed and largely unstructured system of both monogamous marriage and multiple marriage, including both polygynous and polyandrous relationships. Even within this one society, we see a striking degree of diversity in mating systems coexisting without judgmentalism.

In southern Brazil lives another fascinating tribe known as the Pirahã.[8] There is much to say about the unique features of the Pirahã culture, language, and society, and they may be the most famous of all traditional tribes among anthropologists. Their language is incredibly simple. It is tonal, with no important consonants and just a couple of vowel sounds, and it can be hummed or whistled as easily as it can be spoken. They appear to have no words for numbers, colors, or even relatives beyond parents and siblings, nor any religiosity, superstition, or belief in the supernatural. They appear to believe only what they can see with their eyes and hear with their ears. Even stranger (to us), they do not speak of, or appear to even think about, the past or the future. They live purely in the present and are altogether uninterested in anything that isn't part of their daily work of subsistence, recreation, and social interaction.

The Pirahã also have a purely egalitarian social structure in which it is strictly taboo for anyone to tell anyone else what to do or not do. There are no leaders or elders; no currency; they keep no possessions except for basic tools, food, clothing, and some simple jewelry; and they fashion only very simple huts. And, as I'm sure you're probably suspecting by now, the Pirahã have absolutely no hang-ups whatsoever about sex. There is nothing akin to marriage and no word (or concept) for "chastity." Sex is shared freely among members of a community, with no norms or boundaries around gender, age (after puberty), or even relatedness. And when someone rejects another's advances, they can sometimes be convinced by an offer of food, tools, or jewelry. A particularly desirable or skilled sexual partner can even make a living at it.

Moving to Asia, we find a variety of customs and norms, some of which seem bizarre and downright shocking to Western sensibilities. For example, in the Kreung tribe in northeastern Cambodia, adolescent girls are not only allowed but *encouraged* to take their time and explore romantic and sexual relationships with many young men before settling down.[9] In fact, when a young Kreung girl comes of age, her parents will build her a small hut to be used exclusively for her dating life. This is not a dwelling—she continues to reside with her parents until she marries—but rather a sort of "love nest" in which she can enjoy privacy with her lovers. Over the course of several years, she will date many men, often simultaneously, before choosing "the one."

This is not only a practice of sexual liberation, but also of female empowerment, even privilege. Young Kreung men do not get their own love hut, and the decision of when and whom to marry lies strictly with the women. Unmarried men are constantly trying to court and impress women to choose them, and competing overlapping courtships can go on for years. These men must try very hard to appear cool and collected because desperation is decidedly unattractive in Kreung culture. Sexual satisfaction and compatibility are important to Kreung women, and the fruits of fully empowering young women to make their own romantic, sexual, and marriage decisions speak for themselves: both rape and divorce are exceedingly rare and almost unheard of.

The people of the Trobriand Islands, an archipelago off the coast of New Guinea, have completely divorced sex from marriage, if you'll pardon the pun.[10] This tribe is famous for its precocious sexuality. Children begin experimenting and having sex with each other well before puberty, as early as age six or seven. Both boys and girls initiate sex, and sexual relationships are fluid and transient. Sex carries no taboo, is performed openly, and parents vocally encourage their young children to have sex as early as possible.

The taboos that we often find associated with sex in most cultures are instead attached to eating among the Trobriand peoples. Eating is always done alone, usually at home, and with one's back turned toward any others who may be present. It is considered very rude to eat in front of others, and those who are spotted eating suffer embarrassment.

When it comes time for selecting a mate, Trobriand courtship is pretty standard. Couples spend time together talking and having fun; they have sex, of course, and spend the night with each other. To signal their interest in marriage, the woman stays in the man's home after sunrise, instead of scurrying home before breakfast, as usually expected. If the young man's mother accepts her son's choice of bride, she brings them a cooked breakfast and—*gasp!*—they eat together. The sharing of a meal together signifies the consummation of the marriage, and the married couple will eat together for about a year. At that time, the marriage either dissolves or is declared permanent, but either way, they go back to eating separately.

Because sex is shared so freely, and because yams, which have contraceptive properties and can lower fertility, are such a major part of the Trobriand diet, their culture never established the causal connection between sex and pregnancy.[11] Instead, they believe that women become pregnant through the actions of a sacred spirit. Although marriages are established for companionship and to raise children within a family unit, the paternity of the husband is far from certain. In fact, the Trobriand don't have a concept for paternity, and notions of descent and ancestry follow maternal lineages only. Thus the Trobriand have a system of distributed paternity and alloparenting.

Speaking of New Guinea, this mega island is one of the most distinct anthropological environments in the world. It has been inhabited for at least fifty thousand years, populated initially by the same migrants that became the Indigenous people of Australia, which was continuous with New Guinea as one large landmass at that time. Mountainous and densely forested, the nearly impenetrable nature of the New Guinea landscape fragmented its native peoples into countless distinct populations. This one island harbors over one thousand languages, most of them mutually unintelligible to all the others, and some of which are spoken by only a tiny handful of native speakers. A remarkable example of cultural radiation and diversification, there are hundreds of ethnic groups, each with their own cultural practices. The richness of New Guinea's anthropology has been slowly fading since industrialization began with the arrival of British explorers in the nineteenth century.

With such incredible cultural diversity, it is difficult to summarize the sexual practices of New Guinea, but suffice it to say that sexually exclusive marriage is rare. However, many tribes do have an asymmetry in that women are sometimes expected to be faithful, while men are not. In general, there is a common pattern of gender inequality and patriarchy, likely stemming from the markedly belligerent cultures that dominate the island. Because of the harsh life, most tribes are hostile to other tribes and aggressively jockey for prized land areas. Prior to the forced cessation of hostilities by European settlers, many

Guinean tribes existed in a constant state of war with their neighbors, a cultural environment that is always associated with aggressive male dominance.

One of the most peculiar practices found in New Guinea is that of the Simbari people, found in the Eastern Highlands.[12] Like most cultures, the Simbari have a coming-of-age ritual for pubescent boys, but theirs lasts years and is more like a cultural bootcamp. At around age nine, young boys are taken from their mothers and spend several years in the presence of men only, where they learn how to live and behave as a Simbari man. The Simbari have strict gender roles and the adults live mostly as gender-segregated. When the young boys first arrive in the male camp, they are subjected to bloodletting, which ostensibly removes the mothers' presence from their body. Next, they perform fellatio on the adult men of the tribe, consuming semen repeatedly over months or years of time. This is because the Simbari believe that boys are not born with semen but instead must imbibe it from other men. Following this, they are taught how to have sex with women, mostly by practicing on each other.

After a few years, young Simbari men will marry a young woman who is selected for them by the tribal leaders. However, the strict gender segregation persists even through marriage, and men consort with women for sex and procreation only. There is very little contact between adult men and women beyond that, so men typically only have heterosexual intercourse with their wives. Additionally, most adult men do not have sex with each other. Instead, their sexual gratification comes from the steady supply of fellatio provided by whichever young boys are at the right age.

In the Cook Islands, smack in the center of the Pacific Ocean, we find a Polynesian tribe called the Mangaia, described by anthropologist Donald Marshall as "the most sexually active culture on record," with most individuals having sex more than once per day, on average, and with multiple partners.[13] Children of both sexes are taught to masturbate at age eight or nine, and by their teenage years they are sexually prolific. In a testament to how important sex is in this culture, when a boy turns thirteen, he participates in a coming-of-age ritual that involves his having sex with older women from the tribe. Among other things, this aspect of the ritual is meant to educate the young boy on how to have sex properly and ensure that the woman receives as much gratification as he does.

•

We could continue through society after society, but instead we can skip to the conclusion: the notion that monogamous, permanent, heterosexual marriage is some kind of human universal is spectacularly wrong.[14] What I have chosen to include here is just a sample, a smattering of examples that I find curious and informative. Cultures and societies around the world exhibit a wildly diverse array of mating strategies and sexual relationships. There is no "natural state." Every culture seems to do things their own way. Don't believe me? Consider reading one or more of the sources in the recommended reading list at the end of this book, and I recommend starting with *Sex at Dawn,* which explains many of the examples I list above in much more detail. What you will find is example after example of cultures around the world, both contemporary and historical, doing things their own way.[15, 16]

So what *can* we learn from all the various ways that human societies approach their sexual and reproductive relationships? There are definitely some trends that emerge, and one of them is that the local ecology has a big impact. Specifically, when resources are scarce or easily monopolized, there tends to be more social control over sexuality. And when resources are abundant and accessible to all, freedom and diversity flourish.

For example, the harsh climate of the Himalayan mountains or the Sahara Desert means that the populations there subside on the knife's edge of survival. Accordingly, we see more cultural strictures around sex and reproduction, as individuals seek to protect their precious reproductive investments in an unforgiving environment. In contrast, in the lush tropical environment of the Amazon basin or the Polynesian islands, populations tend to be more freewheeling and tolerant of each other's reproductive behaviors. When there is plenty of food, water, and shelter from harsh weather and predators, individuals are much less pitted against one another, and sexual tolerance and diversity follow. But resource scarcity turns survival into a zero-sum game, which stokes competition, rivalry, and attempts to manipulate and control. That's the pattern I see.

This brings us back to the question we asked at the beginning of the chapter—How did sexually exclusive heterosexual marriage became so dominant?—and leaves us with the apparent contradiction that the invention of agriculture seems to have initiated the ascendance of social control over sexuality. Since agriculture led to abundance and the accumulation of wealth, shouldn't it have led to a flourishing of freedom and diversity? If resource abundance breeds tolerance and individual autonomy, why did agriculture and civilization bring the exact opposite in every place that it was introduced?

The answer is power. The plant- and animal-based foods, clothing, and other materials that agriculture produces in great abundance are *owned by someone*. These resources are not taken freely from the environment but, rather, created through directed human effort, which can be organized, controlled, and monopolized. Suddenly, individuals could *own land*—an unheard-of concept among hunter-gatherers—and use it to become wealthy. And this is exactly what happened.

As foraging societies became farming societies, wealth inequality emerged. For the first time, there were haves and have-nots, and the relatively flat and egalitarian social structure gave way to a hierarchical system with power concentrated in the hands of a very few. The prevailing life goal quickly became the maintenance of wealth and power. And voilà, we got the enforcement of "order" and social control. This doesn't necessarily mean a shift to monogamy. In fact, polygyny was even more common than monogamy in agrarian societies all the way through the premodern era. But the point is the same: concentrated power and control brought an end to the era where everyone was free to do as they pleased.

It also appears that it is women who suffered the most during the march toward industrialized and stratified society. In a recent article about the Yanomami, the Amazonian foraging tribe mentioned before, anthropologists Sharon Tiffany and Kathleen Adams had this to say: "Imagine a society in which one woman in every three is raped, usually by a man she knows . . . where one-third of all women are beaten during pregnancy . . . where 20 percent of adult women . . . have experienced sexual abuse or assault [as children]." They weren't talking about the Yanomami. Those statistics were taken from the American Medical Association and describe the situation for women in the United States of America. The Yanomami society would never tolerate such an awful state of affairs.

But another, simpler lesson from all of this is the same truth that keeps bubbling up in chapter after chapter of our exploration: anything and everything we examine in nature looks more diverse and more complex the closer we look. This is because the natural world is a constantly churning, diversity-generating machine, and the natural history of sex reflects that.

(Pre-)Modern Love

The age of agriculture and the reorganization of human groups into towns and cities brought about a whole new way in which humans could interact with one another. What was previously accomplished by gossip, reputation, and coercion was now done with laws, strongmen, and lethal force. The rich and powerful enforced a rigid social order designed to maintain their control and dominance. Like clouds of dust and rock coalescing into stars and planets, a rich landscape of lifestyles, cultures, languages, and customs were condensed and absorbed into megacultures we know as city-states, which themselves unified into nations and eventually empires. A great deal of diversity was lost in the march of civilization, including and especially sexual diversity.

But even from the sparse relics and artifacts that we have from the ancient world, it is clear that sexual diversity persisted and, at least at times, thrived.

The oldest known evidence of a same-sex couple in antiquity comes from ancient Egypt.[16] Two royal servants named Khnumhotep and Niankhkhnum, both men, were entombed together more than 4,500 years ago. Their tomb, not far from Giza, is in the royal burial ground of that era, evidence of their favored status in the royal household. Their position in the hierarchy is somewhat unclear, with different texts and different translations indicating they were, variably, manicurists, hairdressers (because of course they were), heads of wardrobe, and one text most often being translated as "confidants," assumedly for the pharaoh. Although far from universal, it was not unusual in ancient Egypt for highly favored servants and advisers to get their own tombs in the royal burial ground at Saqqara.

What *is* unusual about the tomb of Khnumhotep and Niankhkhnum is that the two lovers were buried together—and clearly depicted as a couple, holding hands, kissing, embracing—despite both of them having wives and children who did not join them in their final repose. In keeping with the grand tradition of the academic erasure of sexual diversity, various scholars have described their relationships as brothers, twins, dear friends, and—I'm not kidding—roommates.

But the evidence that they were lovers is overwhelming. Stunning frescoes adorn the walls of the various rooms and passageways of the small but elaborate tomb structure, and the depictions of the two men get more intimate toward the inner chamber where the sarcophagi once lay. (The sarcophagi and all the grave goods were plundered at some point in the past forty-five centuries.) In the outer passageways, the men are depicted with their respective families, seated with them at royal banquets, etc. Going farther into the tomb, the men are depicted together, without their families, doing a mix of grandiose and everyday things together, from fishing and playing a sport together, to being honored in a royal ceremony re-created in statues with full regalia and accolades. No aspect of how the two men are depicted together is consistent with how siblings, close friends or allies, or roommates (eye roll) are shown throughout ancient Egypt. In fact, in the entire necropolis at Saqqara, these are the only depictions of two men kissing.

Same-sex couples come up a few times in Egyptian relics, but interpretations of the text are somewhat complicated by the fact that sexual intercourse was usually described with euphemisms or flowery language by the ancient Egyptians.[17] For example, King Pepi II had a favorite military officer whom he would visit in the middle of the night to *do as he desires*.[18] In all other contexts, this phrasing is assumed to mean that they have sex, but of course, scholars have offered other explanations, including that it is a metaphoric reference to epic poems and stories. Odd that they never question the "real meaning" when it is heterosexual sex being implied.

Traditionally, scholars have claimed that the lack of mythology or epic poems involving queer love stories or unambiguous records of same-sex marriages is evidence that sexual diversity was not tolerated in ancient Egypt. But the absence of evidence is not the same as evidence of absence. Marriage itself was different then and, until the time of the New Kingdom, was mostly arranged by families to gain or maintain status and promote community cohesion and flourishing. But speaking of absence of evidence, in all of the many laws and codes of conduct, we do not see any proscriptions *against* same-sex sexuality. We do not find sexual taboos at all except for rape. If homosexual conduct was reviled or taboo, it would show up in the law books, as it did in the Hebrew scriptures, written contemporaneously with much of what we have from ancient Egypt.

And speaking of the Hebrews, the liberal sexual attitudes of ancient Egypt are alluded to in the Talmud and related Jewish texts. Egyptians were reviled for many of their sexual practices, including polyandry, a woman having multiple husbands, and lesbianism, which was disparagingly called "the acts of Egypt" by the twelfth-century Torah scholar Maimonides.[19] It's rather telling that polyandry was considered an abomination by a society once led by a king who, according to their own texts, had seven hundred wives and three hundred concubines. It seems that multiple marriage was only revolting when women did it, which is consistent with the patriarchal (misogynistic, if we're being honest) flavor of Judeo-Christian-Islamic morals and culture.

It's important to remember that the restrictive sexual morals of ancient Israel were at odds with those of basically all of its neighbors. According to the Old Testament, the Egyptians were joined by the Canaanites, Babylonians, Philistines, Phoenicians, Ammonites, and others in their ungodly approach to sex. Of course, we can chalk some of this up to the general animosity that rival societies direct at each other, and the desire to stoke revulsion toward one's enemies, but the writings of these societies themselves largely confirm that pretty much all of their neighbors had a more relaxed and tolerant attitude toward sex than did the Israelites.

For example, we know that, while infidelity was a scandal, there was no taboo on premarital sex in ancient Egypt. Both women and men had sex with each other as much as they wanted during their bachelor years. The same was true in Babylon and, further, we have concrete evidence that gay sex was perfectly acceptable in Babylonian culture: the ancient Mesopotamian text *The Almanac of Incantations* contains blessings for both opposite-sex and same-sex relationships. In fact, the only scandal that would accompany a same-sex pairing is if the two lovers were of different social rank, but the same was true for heterosexual pairings, and this is a theme we see throughout the ancient world. No one seemed to care about the sex of the person you had sex with, but sleeping above or below your station—*that* was the real disgrace. I'd say not much has changed on that front over the last three thousand years.

Marriage, on the other hand, was strictly tied to procreation in ancient Egypt and was thus a heterosexual affair, but from what we can tell, this was for the purpose of family making, and no restrictions were placed on outside sexual activity. That's the general theme of most of the societies in the region at that time. It is very interesting to imagine how history might have played out differently if the sexual morals of Babylon, Canaan, or Egypt had become dominant in Western culture, rather than the highly restrictive sexual taboos of the early European Christians, which were even more sex-negative than what they inherited from the Israelites.

Because so much has been written about the same-sex sexual customs of ancient Greece, I need not shed much ink here. Suffice it to say that sexual norms are usually pretty peculiar to a certain time and place, and they are considered curious or problematic elsewhere. When it comes to ancient Greece, that is an understatement because the practice I am now going to discuss, officially called *pederasty*, is that of an ongoing sexual relationship between a middle-aged man and an adolescent boy.[20] While any imagining of this turns our modern stomachs, the practice was not just common but essentially *universal* at certain times in ancient Greece. And the practice was not limited to Athens but was prevalent throughout the Hellenic world, from Sparta in the south to Troy in the east. Such familiar figures as Socrates and Plato engaged in it, either as boys or older men, and probably both, and the practice continued sporadically for many centuries, possibly for up to a millennium.[21]

It is also important to draw a distinction between pedophilia and ancient Greek pederasty. Pedophilia, by definition, is attraction to *prepubescent* children, whereas the pederastic practice in Greece began as a boy *began* puberty. That may seem like a small distinction, but remember that the age at which children were considered mature—the age of consent—was much younger in earlier times. Marriages were often arranged among preteens, and young women often started having children by their midteens. Bible scholars agree that Mary gave birth to Jesus at age thirteen or fourteen—and remember that she was already betrothed by that time. Also, the pederastic relationship was not necessarily consummated sexually right away. Forcible rape, including and especially against children, was legally prohibited in ancient Greece.

The pederastic relationship was about much more than sex. It was a mentoring relationship, and the tutelage covered everything about life that they don't teach you in school. It served as an important coming-of-age practice, a milestone that has specific rituals associated with it in most cultures, and it marked the departure of a young man from the care and protection of his natal family and his entrance into the world as a man. The pederastic "education" included sexual matters so that the young man would be prepared to be a good husband to his wife. It wasn't homosexuality as we commonly understand it.

Just so we're clear, my attempt to understand and explain this practice within its full social and historical context should not be construed as moral approval. Just as with rape, slavery, and genocide, neither the legality nor prevalence of an action is the sole determinant of its morality.

Looking westward, the Romans inherited a lot from the Greeks but also brought their own distinctly Roman flavor to sexual mores. For one thing, restraint and discipline were revered as among the highest virtues in Roman society, and this extended to sex and sexual pleasure. It was not that the Romans were a less sexualized culture than the Greeks, but rather that succumbing to sexual urges was often seen as a sign of moral weakness. Within the approved context, sex was enthusiastically celebrated, especially in pagan times, as fertility and procreation were seen as being essential to the flourishing of Roman civilization.

Although the Greeks were innovators of bilateral marriage, it was the Romans who first elevated marriage to the social bedrock we know today.[22] Surrounded by less "civilized" societies, classical Rome was unique in its commitment to monogamous marriage. Saint Augustine, writing in the fifth century CE, referred to monogamy as a "Roman custom," which distinguished the empire from the surrounding lands in Europe, North Africa, and the Middle East in which polygyny was the norm. This makes sense because marriage in the classical period was much more about acquiring and maintaining status, property, and power for the family, the *house*. The Romans invented the concept of *legitimate* children in order to facilitate the orderly inheritance of property and titles. The romantic form of marriage—love marriages, as we understand them today—didn't truly blossom in Europe until more than a thousand years after the fall of Rome.

But it's important to remember that monogamy and marital fidelity didn't mean to Romans what they mean to us now. This was economic and social monogamy, not sexual monogamy. Marriage was not considered a prerequisite to sex, nor was sex limited to marriage, officially or unofficially, at least not for men. Prostitutes and concubines were abundant and legally sanctioned, and there was no taboo against married men visiting them, although unmarried cohabitation was forbidden. It was illegal, however, for a man to have sex with *another man's wife*, as the Romans referred to married women. This male-centric view of women and marriage underscored the larger sense that property and legitimacy were the key aspects of the marital union.

Male homosexual sex was also part and parcel of Roman society, and it is depicted in graphic detail on literally thousands of pieces of recovered art, from spectacular frescoes to common household pottery. From emperors to commoners, sex between men was, as far as we can tell, free of taboo for the entire republican era and most of the early imperial period as well, until the rise of Christianity.* The exploits of the emperor Caligula became the stuff of legend, but several other emperors are known to have had sex openly with other men, at least occasionally, including Nero, Julius Caesar, and Augustus. The emperor Galba is believed to have been exclusively homosexual.

However, the permissibility of gay sex notwithstanding, the Romans did have some sexual hang-ups. It was a repressively patriarchal society, so rights and citizenship were restricted to men. And while it was perfectly acceptable, even expected, for a Roman citizen to pursue sex with both men and women, this only applied to the penetrative role. Rome saw itself as a society

---

* A curious apocrypha has emerged, mostly among conservative Christians, claiming homosexuality was somehow instrumental in the fall of Rome. Despite being supported by no evidence whatsoever, this supposed fact is often repeated by otherwise well-educated individuals, including the current Speaker of the US House of Representatives, Mike Johnson (R-LA). A simple look at the timeline renders this theory absurd. While it is indisputably true that pre-Christian Rome, both in religion and culture, had much less restrictive sexual morals than did Christian Rome, the empire reached the zenith of its power, influence, and geographic reach around 117 CE, during the reign of Trajan, while the Christian Church was still minuscule and underground. The Western Roman Empire officially collapsed in 475 CE, less than a hundred years after Christianity was made the state religion with the Edict of Constantinople (380 CE). In short, the empire grew and reached its peak while it was pagan. It declined and fell within a few generations after becoming Christian.

of conquerors, and Roman versions of masculinity reflected that accordingly. Being the penetrative partner was akin to a sexual conquest, but being the receptive partner, whether in anal or vaginal sex, was seen as too submissive and thus unbecoming of a Roman citizen. Therefore, the only men with whom a Roman man could have sex (without defiling his partner) were slaves, foreigners, and prostitutes.

This too seems to have varied over time, as a great deal of artwork recovered from the buried cities of Pompeii and Herculaneum (79 CE) nonchalantly depict a variety of sex acts, including a threesome in which one man is penetrating another man who is penetrating a woman.[23] Incidentally, Pompeii also yielded one of the precious few depictions of lesbian sex that have been recovered from ancient Rome. The ubiquity and abundance of various sex acts, including fellatio, cunnilingus, prostitution, and even orgies in the art recovered from these two cities speaks to the much more free and open sexual culture of ancient Rome, aggressively suppressed and erased by subsequent Roman and Italian cultures under the influence of Christianity.[24] Although it took a catastrophic volcanic eruption to protect the evidence from puritanical destruction, we now know quite a bit about the comparatively sex-positive culture of the pre-Christian Roman Empire.*

In summary, the three most foundational classical cultures of Western civilization—Egypt, Greece, and Rome—were all less sexually restrictive and repressive than we are in the modern era. There was no taboo, let alone criminalization, associated with homosexuality, premarital sex, or nonpenetrative sex. Those laws and taboos came much later. While there was a taboo against adultery, at times with criminal consequences, this was defined narrowly as a married woman having sex with a man who was not her husband. Tellingly, we encounter this misogynistic double standard—both officially and

---

* The repression of ancient Roman sexuality was alive and well when the remains of Pompeii and Herculaneum were discovered nearly seventeen hundred years later. After King Francis of Naples stumbled into an exhibition of erotic artwork while touring the Naples National Archaeological Museum with his wife and daughter in 1819, he ordered that all of the obscene artifacts be locked away with tightly restricted access, lest unsuspecting members of the public be scandalized and corrupted. The so-called Secret Museum in Naples kept most of the sexual artifacts locked up, and it wasn't permanently reopened to the public until the year 2000. To this day, minors must be accompanied by a guardian.

unofficially—throughout the world, but always as a function of the concept of *legitimacy* when it comes to inheritance.

Married men were always free to have sex with whomever they wanted, if quietly. But married women had no such freedom because of the doubt this would inject into the paternity of her issue. The primary concern of marriage was not love or even sex. It was inheritance.

Love at Lugu Lake

We are now ready to move our discussion of human sexual practices into modern times. While the steady march of concentrated power and organized religion has stamped out a lot of sexual diversity, it doesn't take much looking around to see that diversity simmering under the surface, ready to reappear when the culture allows.

When it comes to sexual liberation, there is one modern society that deserves its own section in this chapter, and that is the Mosuo people of southern China.[25] Although lumped in with the Nashi ethnic minority by the Chinese government, the Mosuo are a culturally distinct people that have lived in a small region around Lugu Lake for thousands of years. Because of the barrage of tourists and the harassment of disapproving government authorities, their numbers have dwindled to about forty thousand currently. However, the remote and mountainous nature of the region has allowed the Mosuo to resist being absorbed into the larger societies nearby.

The Mosuo have fascinated Western anthropologists since the days when European merchants first encountered them on the Silk Road more than seven hundred and fifty years ago. Marco Polo himself met the Mosuo and was shocked by their sexual licentiousness. He wrote, in 1265:

> They do not consider it objectionable for a foreigner, or any other man, to have his way with their wives, daughters, sisters, or any other women in their home. They consider it a great benefit, in fact, saying that their gods and idols will be disposed in their favor and offer them marital goods in great abundance.

Although what he wrote was correct, Polo was the first in a long line of foreigners to totally misunderstand the Mosuo. Their society is relaxed, peaceful, and agrarian; and, most shocking of all, they have absolutely no sexual taboos whatsoever. Sexual autonomy is absolute, and the only thing that is shameful regarding sex is if someone attempts to restrict someone else's freedom thereof. That said, they do practice an extreme form of discretion: discussion of sex and romance in a *specific* sense is forbidden, as the Mosuo value their privacy as much as their freedom. Discussion of love and sex in the *general* sense is perfectly fine, and Mosuo children are not raised to think there is anything dirty, shameful, or improper about sex.

The Mosuo are usually described as matriarchal, but egalitarian and gender-equal are probably better descriptors. Or, as anthropologist Peggy Sanday puts it, perhaps the kind of society that is built when women are in charge, rather than putting women above men, places the sexes on an equal footing.[26] That said, there are certain areas where women definitely have a privileged position. For one thing, Mosuo families are said to be matrilineal— that is, the cultural and financial line of descendance follows females, who inherit homes and property from their parents. Women are considered the heads of household and make all of the business decisions. However, political positions in the community tend to be held by men, but these positions bring status more than power, as there is very little centralized authority in Mosuo culture.

Mosuo women are fully liberated sexually. When they come of age, they are given a bedroom suite for themselves within the matriarchal household, but this room has its own door to the outside, so that she may conduct her affairs with privacy. Romantic liaisons occur at night, in the woman's room, and the men must leave before sunrise. Relationships can last a single night or an entire lifetime, with most somewhere in between. Because Mosuo women tend not to engage in multiple affairs simultaneously, their relationships can accurately be described as serial monogamy, and because of this, the paternity of most Mosuo children is known. But it makes little difference if it is not.

Children born to Mosuo women are raised in the matriarchal household and cared for primarily by their mothers, with assistance from uncles, aunts, and whoever else shares the home. When the father is known, he may choose to visit his child, but only at night and in the mother's private room. If she has moved on to another relationship, this is not practical and so the father's involvement would be limited to holidays and ceremonial occasions. The Mosuo word for romantic relationships—*axia*—is usually translated as "walking marriage" in English, but the Mosuo strongly reject this because the Mosuo family structure does not involve marriage. Both men and women remain in the household into which they are born, which is run by the women and led by the matriarch. If a father wishes to support his children with gifts, he may, and this may earn him a good reputation, but that is as far it goes. The child will forever be a part of their mother's matriarchal household.

With women not only fully liberated sexually but also empowered within the household, Mosuo men are often considered subservient in our modern, patriarchal understanding of power dynamics. But another way of looking at it—the Mosuo way—is to view the roles of men and women as simply different. Furthermore, under the matriarchal system, the men have it pretty good. Their days are much more relaxed than the women's and their responsibilities are more limited. They help care for the children and the general running of the household, and the better they do with that, the more favor they enjoy from the matriarch, but there aren't direct consequences even if they contribute very little. They are free to pursue whatever *axia* they want, without expectations of commitment. They can be involved in their children's lives if they wish, at the pleasure of their children's mothers, and they raise their own social standing when they are. But no one seems to care if they don't lift a finger. In sum, men can pretty much do whatever they want.

Mosuo women, however, bear the burden of a lot of cultural expectations. Yes, they are free to take lovers with freedom and privacy, but only at nighttime. During the day, until they have children of their own, they are expected to perform household duties at the direction of their mother or grandmother. Mosuo women spend the majority of their days cooking, cleaning, sewing, and shopping, in addition to working in the family business, while the men are free to loaf around if they so choose. And while marriage doesn't exist, it is the duty of Mosuo women to support the growth of the family by bearing children, and she will be the primary caregiver of the children that she has.

The lack of concern about paternity seems also to be connected to a very positive regard for adoption. In most cultures in which bearing children is highly prized, women who are unable to conceive suffer painful shame and stigma, but not so with the Mosuo. If a woman fails to become pregnant after years of trying, her matriarch will arrange for an adoption from a neighboring family. Most adopted children come from families that are either too large to accommodate another child, or are preparing to dissolve because they are too small to sustain themselves (which is becoming increasingly common as the Mosuo population declines). An adopted person is immediately considered fully part of the family and is listed in the genealogy without qualification. The biological parentage is of no concern, so much so that an adopted daughter has just as much chance of becoming the matriarch of her family as a biological one. As an adoptive father myself, it is hard not to see this aspect of their culture as inarguably superior to our own.

Let us now return to the quote from Marco Polo. Notice that the sexual promiscuity is only part of what bothers him. What seems most outrageous to him is that the men are okay with all of this. In fact, the entire passage is really about what is happening to the men. Sex with a foreigner is described as "having his way with . . ." *His* way, the man's way. The women are described as *their* sisters, *their* daughters—that is, the men's sisters and daughters, as if they own them. He also erroneously says *their* wives, because a society where a woman doesn't belong to a man was inconceivable to Polo. It never even occurred to him that women have agency in this culture, are the heads of household, and have sex with the visiting men because *they want to* (gasp!). Had he been able to remove the blinders of his bias, he might have seen a society in which both men and women are free to pursue sex as much as they want, and with whom they want, while still maintaining strong families, cohesive communities, and a vibrant culture.

In summary, although Mosuo matriarchs hold the levers of power within the household, there is little centralized power beyond that. Because men in most cultures are accustomed to bearing the power and privileges of higher status, the empowerment of women feels like the subservience of men. But the reality is that Mosuo women are burdened with more daily work and so-cietal expectations for motherhood, while men are much more free to spend their days and nights however they wish.* Indeed, while many Mosuo men contribute to the homes as doting uncles, sons, and brothers, for others, their lives can be compared to that of a prized stud horse, having plenty of sex but doing little else for their community.[27] The sexual liberation of women equals the sexual liberation of men, and the domestic empowerment of women within households relieves some of the burdensome expectations of men. As feminists have been telling us all along: feminism is good for men as well.

While their overall lifestyle is more "traditional" and low-tech than ours, the Mosuo are a fully modern society, not a foraging one, and they are in frequent contact with their more culturally mainstream neighbors. It has taken concerted resistance, coupled with isolating geography, to avoid being absorbed into the surrounding society, but they are committed to doing just that. And who can blame them? In terms of love and relationships, gender equality, and family cohesion, many would see their society as a utopia, and that is exactly how the Mosuo see themselves. The Mosuo are a contemporary example of a more relaxed and natural state of love and sexuality, showing us how pockets of liberation have always existed even within our globalized modern world.

My Brothers' Keepers

The Mosuo are not the only modern society that has managed to retain a (non-)traditional family life. High up in the Himalayas, small villages dot the harsh mountain landscape. Arable land suitable for growing crops is extremely limited, as is the local fauna of wild animals for food and clothing. But humans always find a way to survive and thrive even in the most unhospitable of regions. Curiously, many of the villages and cultures in this region practice *fraternal polyandry*, where one woman enters into a marriage with multiple men, specifically an entire set of brothers. It is believed that this practice either

---

*   Hopefully this sounds familiar. In chapter 4, I discussed two monkey species—coppery titis and Barbary macaques—in which the matriarchal structure leads to *equality* of privilege and power between the sexes, rather than female dominance over males.

inspired, or was inspired by, the Hindu legend of the Pandavas—five brother demigods that mutually married the princess Draupadi.

Polyandry is one of the least common mating arrangements among humans (and other primates), with one estimate claiming that it has emerged in just four places in human history.* And yet, *eight* different societies in the Himalayas practice polyandry.[28, 29] These eight societies are of different ethnicities (Tibetan, Han, South Asian), different nationalities (Indian, Nepalese, Chinese), and different religions (various sects of Hinduism and Buddhism), indicating that the commonality of the practice in this region cannot be due simply to an inherited cultural practice. These eight societies are entirely different culturally, and so we must conclude that the practice of polyandry is concentrated there for some other reason.

The reason may be simple: fraternal polyandry makes very good sense in the Himalayas. In such an unforgiving climate, with little workable land and few natural resources, there is simply not enough to be divided up among a large group of siblings each generation. Instead, parents arrange a suitable marriage for their oldest son as he comes of age, and then his younger brothers join in the marriage as they come of age. These brothers then mutually inherit whatever land, domicile, or other resources the family has amassed.

The brothers and their shared wife form a large family of mutual support. The brothers divide up the labor, including most of the household chores, and hand over all of their earnings for the wife to manage. She is very much "in charge" of these homes, in addition to doing most of the direct childcare. And the beauty of the system is that if a husband were to die young—which is not an uncommon occurrence in this dangerous region with almost no access to modern health care—his wife and their children will not have to face life on

---

* This count only applies to *strict* polyandry, meaning it is the sole or primary marriage/family structure in that society. As seen throughout this chapter, societies abound where women *can* take more than one lover or husband in the context of a less rigid structure overall. More important, this count also collapses cultural diversity within a specific geographic region. For example, the eight different societies that practice polyandry in the Himalayas are counted as just one instance, but similar geographic clusterings of cultures with similar marriage practices are not typically collapsed when that practice is monogamy or polygyny. It's almost as if there is some kind of bias among anthropologists that leads to the undercounting of polyandry in human cultures. (Shocking!)

their own. With multiple husbands, there is built-in redundancy and lifelong financial security for all involved. And with the brothers jointly inheriting whatever land, home, or resources from their mother and father-uncles, it does not get divided up into insignificance.

Since no one knows their genetic paternity for sure, the children of these families do not make any distinctions among their fathers. They call them all "Dad," and each paternal relationship takes on its own character. This is called *shared* or *partible paternity*, and until these societies came to the modern understanding of sexual reproduction, biological fatherhood was believed to be shared by the brothers. Even now, there is little emphasis on who is the actual genetic father, and in most cases it is never determined for sure. They just don't care about such things. The father who will have the most influence over a son is the one who teaches him the skills of the profession that the young boy chooses.

Girls, on the other hand, are parented mostly by their mothers, who are constantly on the lookout for a good marriage for their daughters. Even though the women are fully in charge of the money in these families, inheritance is passed down jointly to the sons, so arranging a good marriage for her daughter is a key concern for Himalayan mothers. Though it may not seem desirable to us, it works well in these small mountain villages. The villagers themselves have been interviewed repeatedly by outside reporters and anthropologists and are aware of the uniqueness of their particular family arrangements. Nevertheless, they are quick to point out the many advantages of their system and their satisfaction with it.

Hiding in Plain Sight

In the West, the dominant narrative about sex and marriage has penetrated society pretty completely, but deviant subcultures have always existed in the shadows. From the gay enclaves in New York and San Francisco—the Village and the Castro, respectively—to various nineteenth- and early twentieth-century Utopian communes, people seeking an "alternative lifestyle" have usually been able to find them. Sometimes, these alternatives were hidden in plain sight.

First brought into the public consciousness by journalist Terry Gould in his book *The Lifestyle*, the pioneers of the swinging lifestyle in the United

States were none other than United States servicemen and their wives.[30] During World War II, a swinging subculture emerged on military bases among the married fighter pilots. Prior to deployments and during breaks, pilots on the base would hold swinging parties that could be described as rituals. At these parties, the pilots would have sex with each other's wives. This practice obviously flew under the radar, so to speak, because adultery was still grounds for dismissal from the US military at that time. (It still is, actually, but rarely enforced.) Rumors about the practice circulated for decades but failed to grasp the true nature of the phenomenon. While undoubtedly erotic, these swinging parties were not, at their core, about sexual gratification, at least according to the participants themselves. The real reason behind the practice is far more revealing for our exploration of human sexuality.

In World War II, fighter pilots had one of the most dangerous jobs of all, with estimates of the casualty rates being as high as 25 percent. The highly competitive nature of the position, along with the extreme demands of the job, meant that elite fighter pilots formed tight social bonds unlike any that most civilians would ever experience. In addition, the knowledge that many of them would not be returning home from the war created a very specific environment in which the pilots placed their full trust in each other. The mutual trust and support of these pilots extended to their families, who also shared a special bond forged by the extreme stress and emotional trauma of their situation.

These bonds were so strong that they included the obligation to care for one another's remaining families in the event of their death. According to interviews with these swingers, this was the prime motivation behind the swinging ritual that emerged in this highly precarious environment. Reminiscent of the practice of ancient Greek soldiers sharing sexual experiences with each other in order to build trust and cohesiveness, the shared sexual experience of the swinging World War II pilots underscores the power of sexual activity in creating and strengthening social and emotional bonds.

Apparently, enough of the participants enjoyed the sexual openness enough to continue the practice for years after the war ended. In fact, swinging parties quickly spread to other bases and branches of the military. By the time of the Korean War, a thriving swinging scene could be found on most military bases, and it often moved to the nearby towns so that the tight-knit swinger community could involve civilians as well. Apparently it took facing down one's mortality to lift the taboos around sexual monogamy, but once they were lifted, they stayed that way.

Swinger clubs are now found in pretty much every American city, with the largest cities boasting a wide assortment, each catering to different interests and styles of interaction.[31] In New York City, there are several swinger clubs in every borough. Some of them look and feel like nightclubs, with a cocktail bar and a dance floor, but with long hallways of private rooms for more intimate exchanges. Others are more casual and brightly lit, facilitating mingling and easy conversation. Still others operate like community centers and have nightly social activities such as game nights, speed dating, and painting lessons (models are usually nude, of course). Each club has a different vibe and its own set of activities and customs, attracting a distinct subset of the swinger community. In smaller towns, swingers parties tend to be held in people's homes and the community is even more closely knit. One common rule across different swinger scenes is that single men are not allowed, and we can all imagine why not.

Swinging is, and probably always has been, much more common than most people think. As of 2018, 2.5 percent of Americans identify as swingers, but there is more to say about this statistic.[32] For one thing, it has been relatively stable for as long as data is available. However, the term "swinger" is most closely associated with married heterosexual couples, so LGBTQ swingers and unmarried people will usually not identify with it. Surprisingly, even though consensual nonmonogamy is on the rise overall and is enjoying increasing acceptance, or at least tolerance, the swinger community is relatively stable and covert. To me, this says that married heterosexuals have actually been leading the way on consensual nonmonogamy, but quietly. Indeed, studies have shown that not only are open relationships more common among queer couples, but such couples are also much less private about it with friends and family, whereas heterosexual couples are far more likely to fear stigma and judgment and so they keep their sexcapades private. To each their own!

In the West, we often think of the Muslim world as among the most sexually conservative, but in Sunni Islam, the largest branch of the religion, there is an entire category of marriage that exists so that heterosexual couples can have sex with each other, without the usual entanglements of marriage.[33] So-called *misyar marriages* are contractually and religiously valid but designed to be temporary and subordinate to the more permanent and family-building Sharia marriages. Misyar marriages are most often associated with Muslims who travel for their work and must be separated from their spouse for extended periods. It is even acceptable for the misyar marriages to be kept secret from the primary spouse. Misyar marriages can also be contracted when a woman must stay in her natal household for some reason—for example, to care for an elderly or infirm family member—and so a traditional Sharia marriage isn't possible. Misyar marriages have also been conducted for no other reason than the fact that a man desires sexual variety but does not want to engage in sin nor lead a woman to do so.

Shia Islam is even more direct. *Nikah mut'ah* directly translates to "pleasure marriage" and is a sacramental, but not legal, marriage restricted to a predetermined length of time.[34] Participants can be single or already married, and almost none of the other trappings of marriage are involved. It is purely to allow sexual relations, and the contract term can be measured in minutes. The practice was officially sanctioned by Muhammed himself, but it has become controversial in the Muslim world because of attempts to use this practice to legitimize prostitution, which is forbidden. Many Islamic scholars, however, point out that the key distinction between the standard marriage and the *mut'ah* variety is that the former is explicitly to foster procreation and the creation of a family unit, while the latter is designed only for sexual pleasure. The acknowledgment that these two goals are distinct is a stunning admission of a reality that has eluded most Christian denominations.[35]

Speaking of devout Muslims and adventurous sexual escapades, on the island of Java in Indonesia, there is a sacred mountain called Mount Kemukus. This mountain is home to a shrine to a local saint, who was a prince who abandoned his royal duties and titles in order to run away with his beloved. According to legend, after running away together, the two made their home on the mountain, and it has since become a shrine in honor of their devotion to each other, against the customs of the day and the expectations that society had for them. Considering how strongly social pressure in Indonesia promotes conformity to norms and morals, the commemoration of a rebellion against said norms and morals seems a bit odd, but then again, that's probably the point.

To celebrate the pursuit of love against all odds, the locals celebrate a special festival called Pon every thirty-five days, during which participants have sex, outdoors, with total strangers.[36] Yes, Pon is a celebration of sex and eroticism, but it is also a Muslim service, complete with blessings, invocations, offerings, and morning prayers. At the beginning of the night and again in the morning, participants lay flowers at the grave of the prince and bathe in a sacred stream. Some travelers often come with a specific, nonsexual prayer request for which the pilgrimage is ostensibly dedicated, such as the healing of a sick relative, a promotion at work, financial success, a child to be born healthy, etc. Some travelers are looking for love. Obviously, some are in it just for the sex.

Many, if not most, of the participants are already married, and for them, the festival is simply a sexual affair. But others are there hoping to find sexual excitement and compatibility that will mature into a relationship. In fact, to get the full effect of the pilgrimage and the blessings that are promised to follow, participants must return and have sex with the same partner for seven festivals in a row. That's seven nights having outdoor sex together, in public view, in the hopes of gaining favor and blessings.

Predictably, a cottage tourism industry has sprung up to accommodate the pilgrims, and the festival has expanded to include nonsexual, even family-friendly, elements. The growing popularity of the festival and its steamy nature has meant that the festival organizers have an uneasy relationship with the local community in the village of Solo. The tourism has brought much needed funds for development to a once remote and underdeveloped region, but the nature of the festival offends the sensibilities of most Muslims; Indonesia, and Java in particular, is known for being devout and conservative. Most efforts to ban or curtail the festival have failed because, well, money talks.

These various traditions that exist within a religion that otherwise values chastity, almost above all else, seem puzzling, but what they tell us is that our beliefs and convictions often reside alongside the very human drive for sex and sexual variety.

Open Secrets

Eleanor Roosevelt was probably the most influential First Lady in US history. Serving from 1933 until her husband's death in 1945, Roosevelt did much more than pick out place settings and host cocktail parties. She helped shape policy, gave speeches and press conferences on the president's behalf, and brought many causes to the forefront of the Roosevelt presidency. She was outspoken in her support of the growing civil rights movement for African Americans and advocated for the advancement and protection of women in the workplace. She even publicly disagreed with her husband on occasion, though she was most often a loyal and tireless champion of his New Deal policies. Following the death of President Roosevelt, she remained an active public figure for the remaining seventeen years of her life, including serving as the first US ambassador to the United Nations, a position that still bears the marks of her considerable influence, as does the role of First Lady.

Roosevelt was also almost certainly a lesbian or a bisexual and is believed to have had multiple long-running affairs with both women and men. One relationship stands out as particularly significant for her. Lorena Hickok was the first woman to have a piece in the *New York Times* under her own byline and one of the first great woman reporters in the United States. She covered the Roosevelts for the Associated Press and eventually became a biographer of Eleanor. Although Hickok was just one of many close lesbian friends of Roosevelt, the two spent a great deal of time together, becoming the closest of friends, a friendship that eventually forced Hickok to resign her position with the AP because she could not be objective in her White House coverage.

Despite her grueling schedule, Roosevelt wrote long letters—often dozens of pages—to Hickok almost daily when the two were not together.[37] Although the written record has been extensively "sanitized" by family members, biographers, and Hickok herself, more than *three thousand* letters between the two have been recovered, detailing intense devotion to one another and describing explicit longing for physical affection. At her husband's first inauguration, Roosevelt wore a sapphire ring that Hickok had given to her and, later that day, gave her first interview as First Lady to Hickok in a women's restroom in the White House. That they were in a romantic relationship is no longer in serious dispute among historians, and it also appears that not only was the president aware of the relationship, he was supportive. (He too had many ongoing romantic affairs.) Eleanor had her own cottage on the Roosevelt estate in New Hyde Park, and Hickok stayed there more frequently than all her other guests put together.

The most influential, accomplished, and revered First Lady in United States history had a years-long extramarital relationship. And her husband, perhaps the greatest US president of the twentieth century, was perfectly fine with it.

Many decades ahead of their time, Franklin and Eleanor Roosevelt had what would now be called a modern marriage. Each took lovers and maintained their own bedrooms—sometimes separate dwelling spaces altogether—in each of their homes and estates. They allowed each other a private intimate life that was not subject to the jealous whims or insecurities of the other. And yet, by all accounts, they had a happy life together as well. Contemporary eyewitness accounts are not rife with stories of loud arguments nor cold indifference. They loved each other.

What's more is that Franklin and Eleanor were devoted life partners. While pushing for women's rights and equality, Eleanor also energetically supported her husband's political aspirations and was his chief surrogate on the campaign trail, especially when he was stricken with the debilitating symptoms of polio. Both as New York governor and as president, Franklin turned to Eleanor as his most trusted adviser, almost his co-president. Near the end of his life, Franklin was occasionally incoherent or even unconscious for extended periods of time. Long before the passage of the Twenty-fifth Amendment and in the midst of a world war, constitutional questions were quietly raised about who held the reins of presidential authority while he was incapacitated. Appropriately or not, Eleanor filled the void. She conveyed his agenda and made decisions on his behalf during these episodes and the White House staff followed her lead, knowing that she had the president's full trust. They were the quintessential power couple.

Perhaps the greatest evidence of the strength of their partnership was the very sexual freedom that many would label as infidelity. Franklin had affairs with his secretaries and Eleanor had both male and female lovers, but, as far as the record shows, their marriage remained a strong and enduring partnership, despite Eleanor's initial unease with Franklin's affairs. They may not have had much of a romantic or sexual relationship together, in fact. Eleanor was quoted by her daughter as having disliked sex with Franklin, considering it "an ordeal to be borne." But bear the ordeal she did, having six children over ten years' time. It's not clear if her distaste for marital sex was specific to Franklin or toward men more generally. We'll never know for sure. But how much does the quality of sex really matter in a marriage after forty years, or even after just ten?

The Roosevelt marriage shows us the distinctions between sexual monogamy and social monogamy in humans. While they shared their beds with others, they remained devoted to each other and to the life they had built together. Sexual fidelity is not a requirement for committed and permanent social monogamy. Their success as individuals was tied closely with their success as a couple. And the same could be true of their genetic posterity, which continues its expansion today. The Roosevelt name continues to draw awe and prestige, as it did long before Franklin and Eleanor tied the knot. (Eleanor was actually a distant cousin of Franklin, so she did not have to change her surname when she married.) The two of them together were more than their sum as individuals. Partnership, devotion, companionship, mutual love and support, shared parental investment: these are the true features of monogamy in our species and the key elements of a strong marriage.

Secret Secrets

Karl Koecher grew up in what was then called Czechoslovakia.[38] A complicated, narcissistic, and megalomaniacal man, he eventually dropped his anti-state activities, switched his loyalties, and joined the Communist Party to see if he could advance himself within it. Because of his piercing analytical abilities and his command of several important languages, he was recruited by the Czech security service, which was largely subservient to the Soviets at that time. Karl and his new wife, Hana, were chosen to go to the United States as spies. They spent two years developing their backstories as political dissidents disillusioned with the Communist regime. When the time was right, they fled to Austria and eventually emigrated to the United States. They established themselves in New York's fashionable Upper East Side, were regulars at the famous Studio 54, and became Manhattan socialites climbing the social ladder. Their mission was to penetrate the CIA, and the method they used to do that was "up to them."

As a defector fluent in Czech, Russian, English, and French, Karl easily found work in the US government, first at Radio Free Europe and eventually at the CIA. The couple gained a fragile trust from their superiors, who always wondered where they got their money, and they worked together as a husband-wife team, among the most prolific double agents in the history of the Cold War. They accomplished their mission, successfully infiltrating the highest levels of the CIA and providing invaluable information to the Czech and Soviet authorities. The damage they did to US interests was substantial, but in order to maintain trust, they also fed classified information back to the CIA. At one point, they were on the payroll of the Czech security services, the KGB, *and* the CIA.

None of their handlers fully trusted the Koechers, but all of them benefited from their work consistently enough to keep them on the payroll until the mid-1970s, when the FBI began to surveil the pair, an investigation that was often overlapping with the CIA's own surveillance. The Koechers had a suspicious amount of money and no one fully understood how they were able to move in and out of elite circles so easily. Karl and Hana's specialty was their unique ability to obtain *Kompromat*, a Russian slang term for embarrassing private information about someone that makes them subject to extortion and blackmail. They did this by joining their two great passions: espionage and sex.

Sexpionage, using sex to obtain classified information from, or compromising material about, intelligence agents or high-ranking government officials, is a practice as old as history itself. But the Koechers may have been the most prolific practitioners of it. After honing their craft at sex clubs in New York, they eventually gained entry into the underground swinger scene in Washington, D.C., where they successfully made contact with, and compromised, several officials at the Pentagon, midlevel bureaucrats throughout the government, and more than ten high-ranking CIA agents. Hana even had sex with a US senator, while Karl had sex with the senator's wife.

Jonna Mendez, the former head of disguise for the CIA,* described Karl and Hana Koecher as among the most devastating spies in US history. Loyal only to themselves, this husband-wife spy duo managed to be fully entangled with both the CIA and the KGB—*at the same time*—working for both, lying to both, and providing both with intelligence against the other throughout the Cold War. Not only that, they were not exactly holed up in a dilapidated

---

* Jonna Mendez is also the wife of the late Tony Mendez, a CIA agent played by Ben Affleck in the movie *Argo*.

safe house as clandestine agents. They rode out their cushy life for many years, enjoying fancy luxuries, lavish financial rewards, and high social status along the way. They were hiding in plain sight, and if that wasn't brazen enough, they also got to enjoy a whole lot of sex with important people and their spouses.

All because of our squeamish taboos about sex. Yes, they had language skills and were very intelligent, but what really made Karl and Hana special was that they were open to engaging in consensual nonmonogamy—that is, they were willing to allow each other to have sex with other people. (They were not just willing but eager, it turns out, but that's beside the point.) The secret swinger subculture, and their own sexual reputation within it, opened many doors for them and gave them access to powerful people with secrets. Hana was regarded by all as stunningly beautiful, a classic blonde bombshell with bright blue eyes, long eyelashes, and a shapely figure. Karl was average looking, but he had a reputation of being something of a sexual Don Juan.

The Koechers' ability to attract and seduce powerful people into bed was quite impressive. Funny thing about humans: when our clothes come off, our lips get loose. The Koechers usually didn't have to probe very much or ask many questions before their targets started talking. After all, even intelligence operatives and US senators like to kvetch about work after a long day. They weren't stealing nuclear codes or strategic plans, but valuable intelligence can come from even the vaguest information. Simply knowing that a particular CIA operative was about to leave town, for example, could help the KGB piece together events. Knowing who reports to whom, who is assigned to which project, language and region expertise, work schedules, focus cities, even budget priorities are all things that could innocently slip out during pillow talk and provide valuable insight to adversaries.

In most cases, the even bigger secret was simply the fact that the target was having extramarital or homosexual sex in the first place. Even without saying a word, by having an illicit sexual encounter, a senator or cabinet official was comprising himself and becoming vulnerable to coercion and blackmail. By threatening to leak evidence of their "immoral behavior" to the press, the KGB could turn a target into an asset. And as the relationship continued, the asset was routinely asked to do more and tell more. These reluctant assets lived under the constant threat of exposure of their escalating misdeeds. All because of sexual taboos.

Let's imagine instead a senator who is openly gay and has no need to seek clandestine sexual liaisons to meet his sexual needs. Or an intelligence officer who, together with her husband, enjoys swinger parties and spouse swapping every now and then, always among consenting and sober adults. Or a polyamorous cabinet official who raises children with her chosen nesting partner and has additional romantic relationships as well, and maybe the occasional hookup, without shame or secrecy. We are quick to blame the hapless victims of *Kompromat* who foolishly expose themselves to extortion, but if it weren't for sexual taboos, none of these individuals would be compromising themselves simply by living their lives. And unscrupulous opportunists such as the Koechers would have nothing to hold against them. There is no end to the harm done by taboos regarding what consenting adults do in bed together. And it's totally unnecessary.

Hopefully this whirlwind tour through human sexual diversity served to demonstrate that the recent sociosexual upheaval isn't really anything new or radical. What we're seeing are the convulsions of a highly excitable species whose sexuality has suddenly been unleashed after centuries of repression. Sexual orientation and attraction is a powerful draw that humans have toward one another, and this draw can lead us toward connecting with people in a variety of ways and for a variety of purposes. As with animals, we see that our sexual activity is about much more than simply conceiving children, although it definitely does that. And it's about much more than obtaining pleasure and gratification, although it definitely does that too. Sex is among the most central and ancient human behaviors, and the way that we engage our sexuality is based partially on innate biological drives, and partially on the culture in which those drives take shape. We saw in this chapter a small sample of the ways that human sexuality has taken shape in different times and places. In the final chapter of our journey, we will bring things full circle and return to some of the current social phenomena that we mentioned in the introduction. When it comes to these crazy "postmodern" expressions of sex, gender, and sexuality, everything old is new again.

Chapter 7
The Sex and Gender (Non)Binary

Of all the topics covered in this book, transgender visibility and acceptance seems to unsettle people the most. Even for many progressives, including some gays and lesbians, there is a sense that the transgender liberation movement is going "a little too far" and that maybe not all of those who call themselves trans really are so. It seems to me that this hesitance from progressives stems, at least in part, from the same misunderstanding that drives the very active resistance from conservative reactionaries: that is, from the sense that being transgender isn't really a *biological* reality but rather a cultural or psychological one.

Let me first say that being biological should never be a requirement for considering something real and valid. For example, race is primarily a social construct with almost no basis in biology, but that doesn't mean it isn't a real identity and a distinct lived experience with social, financial, psychological, and legal consequences. The same is true for being wealthy, substance or behavior addicted, having PTSD, being considered attractive (or not). While these may have a biological component, they are primarily psychosocial conditions, and denying their existence because they aren't purely reducible to biology is as absurd as it is harmful.

But sex and gender variations most definitely *are* biological in nature. Think of the cooperator sunfish, the silent crickets, and the dull-colored flycatchers; or the male garter snakes with female pheromones; or Donna, the asexual chimpanzee; or the male gorillas who have left the harem lifestyle behind. These animals do not fit the expected mold of how males and females are supposed to behave, and yet, there they are, living their lives. The behaviors of these animals are biological; it is our *expectations* that are cultural and psychological. How much do you think Donna cares about our expectations about her gendered behaviors? Transgender humans are in keeping with a very long tradition of gender diversity in social animals.

Of course, the gender diversity of humans may not perfectly align with that observed in other animals, but the animal examples that we've explored don't perfectly align with each other either, nor should we expect them to. Each species has its own unique ways of doing things—of doing *everything* in fact. The gibbon and the orangutan are both Asian apes, and yet their approaches to sex, gender, and reproductive behaviors could not be more different, and that's the whole point of this book. It's not just that all species have sex and gender diversity. All species have *their own* forms of that diversity.

Another big stumbling block for transgender acceptance is that most cisgender people, including many who would describe themselves as allies, believe that a transgender person should fully identify with the opposite gender than was assigned at birth; they should fully immerse in all of the accoutrements of that gender—hair, clothing, jewelry, makeup (or lack thereof), mannerisms, vocal inflections, etc. In other words, they should try to "pass" as much as possible and, once the transition is completed, to simply *be* the new gender and move on as though they aren't transgender anymore because the problem has been resolved. Anything less than that makes many people uncomfortable.

But most transgender people don't exist that way. Like many others in the queer community, most trans folks are more committed to *challenging* gender norms than conforming to them. For many, the goal is not to pretend to be cisgender but rather to be authentically themselves by casting away gender restrictions and boundaries altogether and expressing their masculinity and/or femininity in various creative ways. It is precisely that abandonment of society's gender expectations that is so upsetting. It is as though we want to tell trans people, "Okay, you are a girl or you are a boy, which is it?" We prefer hard categories with clear definitions. But at its core, transgenderism teaches us that the categories *aren't* strictly defined or permanent. People can exist at the margins, in between, or in their own sphere.

Isn't that what we see in other animals as well? The cooperator sunfish don't build nests like most males do, but they don't act like females either. Some male salmon don't grow large and don't fight with other males, but they don't mimic females *at all*. The satellite male ruffs look just feminine enough to avoid aggression from other males, but they don't behave like or flock with other females. The "mobbing" male marsh herriers have a mix of male and female physical traits, but they don't compete with males for territories and they *do* cooperate with females to drive out nest predators. Donna is a somewhat androgynous female who acts in many male-typical ways, but she shows no interest in sex with males *or* females.

What we see when animals experiment with gender is not a member of one sex trying to pass as the other sex but rather individuals adopting an atypical combination of masculine and feminine physical and behavioral traits and, in so doing, finding a place within the population. In my view, this more closely matches what I see among transgender people. They are not always trying to "blend in" within a gender but trying to stake out space of their own.

Transitive Properties

I also have noticed that, as the prevalence of people identifying as transgender has increased, so has the skepticism.[1] Probably out of pity masquerading as compassion, many of us are willing to accept a trans person here and there, as long as they are few and far between, meek and isolated. But if the community gets too large or pushy, we start to hear things like "Where are all these trans people coming from? There weren't near this many when I was growing up!"

Some people think of transgenderism as wholly new and aberrant, while others seem okay with supporting trans people as long as they are rare and quiet. We can't have them upsetting our cherished gender norms and expectations. Worse yet, many reactionaries are invoking the term *social contagion* to describe the increase in children identifying as trans. The idea is that if being trans is seen as cool and trendy, it can spread like an infection.

This is an odd position to take when it is crystal clear that being trans is much more likely to get a child bullied at school, rejected by family, and ostracized from their church and other social groups. Most teenagers just want to fit in! But this also raises, I think, an even more fundamental question: How are we to know how large the transgender population truly is without letting people decide for themselves, free of the social pressures to conform?

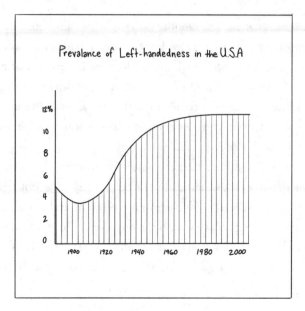

Changing prevalence of left-handedness
in the United States over time

There is a parallel for comparison. The number of people who identified as gay prior to 1950 was vanishingly small. However, as the gay rights movement picked up steam and encouraged closeted gay people to come out, the number of self-identified gay people grew throughout the 1980s and 1990s. Changes in social acceptance didn't *make* people gay, it simply allowed them to be open about who they were and to live authentic lives. The numbers have since leveled off, and the percentage of the population that identifies as homosexual or bisexual seems to be holding steady over the last two decades at around 6 to 8 percent.[2]

Don't like that example? Here is another: around the turn of the twentieth century, only about 3 percent of the population in the United States identified as left-handed. Beginning around 1910, this number began to steadily increase and finally leveled off at around 12 percent in 1960, where it remains today. Did people just want to buck the convention and be cool? It would be absurd to believe that left-handedness is contagious, so what are we to make of this trend?

It's obvious in this case: prior to the twentieth century, left-handedness was not tolerated. Children were *forced* to use their right hand, both at home and in school, often with an enforcement regime that included corporal punishment. Around the turn of the century, we as a society began to see this absurdity for what it was and stopped forcing children to develop right-handedness when their dominant hand was clearly the left one. Some simple accommodations were made, such as the creation of left-handed scissors and desks, and, accordingly, left-handedness began to rise until it reached its "natural prevalence" in the population, around 12 percent.[3] Once the barriers and stigma were removed, the underlying biological reality was revealed.

Gender theorists believe that is what is happening now with transgender people in the human population. For centuries, the barriers and stigma were so strong that vocal expressions of an inner transgender identity were essentially forbidden. And now that transgender people have a chance of affirmation, or at least tolerance, from some parts of their social circle, their identities may now blossom. As this trend levels off, we will have some idea of the "natural prevalence" of transgender people in the human population.[4]

In the previous chapter, I was able to reach back to antiquity for ready examples of homosexuality and nonmonogamy. While it's not as easy to do that with transgenderism, we do have plenty of examples in other cultures and time periods in which gender roles and norms were not as strict as the gender binary we find in most of modern society; and when that binary is not created or enforced, people naturally existed on a spectrum. For example, it has been estimated that, in more than 130 of the recognized Native American tribes and nations, a more expansive and flexible notion of gender was the norm, and it was commonly recognized that not everyone fit neatly into the category of male or female.[5,6] The term "two-spirit" was officially adopted in 1990 by many of the First Nations as an umbrella term to describe the role that transgender people play—and have always played—in their social and ceremonial traditions. Before European settlers bulldozed through the continent enforcing conformity along the way, gender freedom and fluidity reigned through these purple-mountain majesties.[7]

The lack of terminology for transgender individuals in the modern world has created difficulties for seeing them in the ancient, premodern, and even prehistorical eras. Just as biologists missed both same-sex sexuality and extra-pair paternity in animals for all those years, historians and archaeologists failed to see what was plainly visible in ancient texts and artifacts. Remember Khnumhotep and Niankhkhnum, the male lovers from ancient Egypt, buried together with murals and shrines dedicated to their partnership but described by twentieth-century Egyptologists as brothers or roommates? Bias infiltrates every aspect of how we interpret evidence, and we find only what we look for. For example, for almost two millennia, one of the only words we had from ancient Greek to refer to queer people was *arsenokoites*, literally meaning "male beds," though it is usually translated as "homosexual." But this word was never used by the Greeks. It originated with Saint Paul in one of his letters to Timothy and is used solely for condemnation.

Beginning in the middle of the twentieth century, scholars have begun to make progress in uncovering our gay and transgender past. Queer people and stories were hiding in plain sight. We just didn't see it because there wasn't any special note taken of them. As historian Joshua J. Mark states:

> There are not even words in the ancient languages which translate to the modern-day "homosexual" and "heterosexual" which were only coined in 1869 CE . . . Distinctions concerning sexual identity, and prohibitions on same-sex relationships, only begin to appear after the rise of Christianity, which rejected practices associated with earlier religious beliefs.[8]

In other words, there weren't words for "gay" or "transgender" for the same reason that there weren't words for "straight" or "cisgender." It was a distinction that wasn't important at the time. Making matters worse, the ancients did consider sex a mostly private matter and most often wrote about sex euphemistically, just as we do today. But just because they were polite doesn't mean they were trying to conceal anything. Mark continues:

> In ancient Mesopotamia, the priests and priestesses of the popular goddess Inanna (better known as Ishtar) were bisexual and transgender. One of the aspects of the goddess considered most awe-inspiring was her ability to turn men into women and women into men, the power of transformation. Her father-god Enki is said to have created a third gender, "neither

male nor female," who became Inanna's servants, her clergy. What is today referred to as "non-binary" gender was recognized over three thousand years ago as a third gender created by the divine will.[9]

Speaking of hiding in plain sight, one of the earliest and most colorful stories of a transgender person is that of Catalina de Erauso, a seventeenth-century Spanish adventurer who, after escaping life as a nun, began a life of constant travel, exciting employment, and conflict.[10, 11] Their* adventures began in the small Spanish town of San Sebastian, where they spent time living with family, learning Latin, stealing from relatives, absconding with foreign travelers, and at one point obtaining a position as a page in the court of King Philip III. After moving to Bilbao, Erauso spent time in jail for severely injuring a teenage boy who had attacked them. All of that time, since leaving the convent, Erauso presented as a man, going by various names, including Pedro de Orive, Francisco de Loyola, Alonso Díaz Ramirez de Guzmán, and eventually Antonio de Erauso, as they would become known in the New World.

Apparently, their male presentation was so different from their prior female one that they weren't recognized even by close relatives, including an aunt for whom they worked for a time, former convent colleagues where they later attended Mass, their mother's cousin, and even their own father, who came looking for them and spoke with them at length about his "missing daughter." Eventually, they followed the call to America where even more adventures awaited, leaving Spain in 1603.

In Spanish America, their travels took them to present-day Panama, Venezuela, Ecuador, Peru, Argentina, Mexico, and extensive military campaigns in Chile. They robbed and killed their uncle, cut off the face of a rival, spent several stints in prison, owned at least three enslaved Africans, and became known as a conquistador of excessive cruelty to natives while contributing to the conquest and subjugation of the Indigenous peoples of modern-day Chile. They were convicted multiple times of vandalism, were involved in several duels, and murdered many travelers during robberies, almost always escaping justice. They even killed their own brother. Eventually, they received

---

* I use singular-neuter pronouns for Erauso because we have no way of knowing how Erauso identified, let alone which pronouns they preferred.

a long overdue death sentence for multiple murders but, keeping to form, managed to again wiggle free by bribing a fellow inmate into giving a sworn statement exonerating them. It is believed that they later killed the inmate rather than pay the promised bribe. In addition to the extensive military career in Chile, they also spent some quieter periods working with merchants and as a smuggler of wheat and other produce.

Despite being born and raised as a female, and having been sent to a convent at a young age, Erauso lived this rootless life of treachery as a man. They kept their hair short, dressed as a man, and took several female lovers. They would only reveal themselves as anatomically female as a last resort to escape justice, a tactic that usually resulted in confinement in a house of religious worship rather than jail, a much easier place from which to escape, as they always managed to do. Four centuries later, we cannot know if Erauso was a lesbian, a transgender man, or identified in some other way. What we know is that Erauso lived most of their life as a man, despite being born and raised as a girl. That said, their physical aspect was described as masculine enough to pose little trouble in their passing as a man.

As one of the earliest complete accounts of the life of a (possibly) transgender person, the cruel and outlandish adventures of Erauso remind us that transgender people have always existed and somehow found their way among the population. While their character and temperament were monstrously evil, the manner in which Erauso engaged their gender expression in their particular historical context is fascinating, to say the least. At that time in Spain and Spanish America, women had little autonomy, opportunity, or social mobility. But as a man, Erauso rubbed elbows with bishops, nobles, generals, and conquistadors, and their travels even brought them face-to-face with a king and a pope. So, as unique and horrific as the story of Erauso is, it is just one of the millions of stories of transgender individuals creatively crafting an existence in a culture that is hostile to them.

I see a direct parallel between diverse gender expressions in humans and the gender diversity we see in other animals, from sunfish to chimpanzees. Because human society is so highly gendered, the way that we express our masculinity and femininity is key to navigating our lives. Just think about how much our sexual expression can matter on a daily basis. Almost every interaction we have is colored by our gender, from our experiences in the workplace, to roles and expectations in domestic life, to how we are treated on public transportation. For Catalina de Erauso, presenting as a man meant an entirely different life than presenting as a woman.

Being transgender goes way beyond strategic outward presentation, of course. We are not talking about male or female mimicry. Remember, though, that we evolved and adapted in a world that is very different from the one we now inhabit. While the culture and society that surround us are not encoded by our genes, they are indeed "passed on" through social inheritance. And the way that our sex and gender first take shape in our bodies and minds is indeed encoded by our genes, or at least heavily influenced by them, but the social environment affects how the genes do their work. As we saw with the mountain gorillas of Virunga, genes only take shape as behaviors under the influence of environment. When the environment changes, so do the behaviors, within a range established by genetics.

All living things, including humans, are diversity-generating machines. This is true for sex and gender perhaps more than for any other aspects of our physiology. Why *wouldn't* nature play around with our sexual identity as part of the normal process of tinkering that is part and parcel of sexual reproduction? Biological diversity is random variation around a central theme, like the scatter-shot of a shotgun fired at a target far away. The shot that leaves the gun mostly heads toward the target, but it scatters around a central point, drifting farther from that point the longer it flies. Sex and gender is mostly headed toward the target of reproduction, but it is expressed in a spread of diversity in the population, like a scatter-shot. Why? Because in nature, unlike at a firing range, the target is always moving. The environment is always changing.

When it comes to sex and gender, the "environment" that I'm talking about is often purely social, not ecological, and this is true for animals as well as humans. Consider how the social environment of the Virunga mountain gorillas totally changed over the past few decades. And yet they didn't miss a beat. There was no disruptive or precarious upheaval in the social fabric. Having built-in variation for how gender can take shape allowed the gorillas to simply adapt without having to wait around for new mutations to emerge. That kind of flexibility has allowed them to thrive.

For Catalina de Erauso, having a built-in sense of masculinity, or perhaps gender fluidity, allowed them to adapt as well. They were born with gender variation as part of their tool kit to survive and thrive, and the conditions were right to activate that potential. Importantly, if the trick to surviving the sixteenth century as a woman was simply pretending to be a man, then droves of women would have done that. As it happens, there was something different about Erauso, an inner identity that drove an outer expression that allowed them to thrive as a man. We know that Erauso wasn't alone and that history is replete with similar stories of individuals who engaged a gender that was not perfectly in line with their anatomical sex at birth. One of the great things about gender as a social construct is that it can be constructed in a myriad of creative ways. That's the whole point: gender has never been a biological binary.

The Sex (non)Binary

As the idea that gender exists on a spectrum has taken hold on society, the distinction between gender and biological sex has increasingly become the subject of controversy and disagreement, even among those who are not otherwise transphobic or reactionary. Because of the very human tendency to prefer categories and dichotomies, for many, letting go of the gender binary has meant tightening the grip on the sex binary.

Biologists aid and abet this effort when we insist that males make sperm and females make eggs and that's all there is to it. This is a linguistic and scientific trap, and it is altogether unnecessary. But like all traps, the sex binary is tempting. It has the allure of clarity and absolute certainty. If a creature—human or animal—has the organs for making eggs, she is a female. If he has the organs for making sperm, he is a male. Nice, neat, clear. No arguments, no controversy.

Oh, but then there are the hermaphrodites. Individuals that can make both sperm and eggs—either at the same time or in different life stages—are not just common but *the norm* in many different kinds of animals. So, at the very least, there are *three* sexes of animals, rendering the word *binary* inaccurate right from the jump, at least for many kinds of animals.

But biological sex can and should go beyond that. What do we mean when we say that someone is a man? Are we *really* talking about which gametes are made in their gonads? How often is that relevant? Whether we are

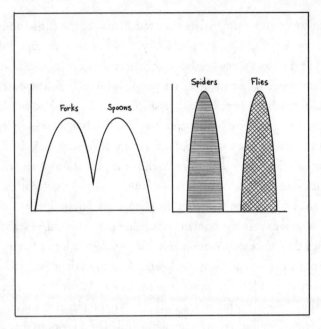

Bimodal versus binary distributions

talking about bathrooms or sports teams, we are never concerned about eggs or sperm. If someone's gender refers to internalized identities and outward expressions, their biological sex refers to their *bodies*. Their *whole* bodies.

The sex of a body is an important matter, biologically speaking. One's sex impacts things like the levels of many hormones and their cyclical patterns, what kinds of diseases they are likely to get, birth control and safer sex—even something as specific as selecting cancer treatments can be impacted by biological sex. And none of these have anything to do with sperm or egg cells, per se. For example, Hashimoto's disease, an autoimmune disease that attacks the thyroid gland, is eight times more common in women than it is in men. Lupus is nine times more common in women. Men, on the other hand, are more likely to have high blood pressure, heart attacks, and congestive heart failure, as well as gout and hemochromatosis (while women often have its opposite, anemia). Men are also far more likely to suffer from addiction and to die by suicide. When a doctor considers your sex when determining the best course for your health care treatments, s/he is thinking about your entire body, not just your gonads.

When we consider the maleness or femaleness of the entire body, we see the inadequacies of using a simple binary.[12] Men are, on average, taller than women, but the size ranges overlap a lot. There are stereotypical body shapes for men and women, but we all know people who don't fit these stereotypes very well. Yes, women have larger breasts that can lactate, but, frankly, so do some men. Yes, men have beards and hairy bodies, but there are plenty who don't—and some women who do, for a variety of reasons including just natural variation. Even if we dial down to the genital anatomy—and that seems to be what some people obsess over—there is a lot more variation and androgyny than is commonly known. In almost any physical measure you can think of, what we see in the human population is two bell curves that overlap. Human sex differences represent a *bimodal distribution*, not a binary one.

Even more important, individuals who don't fit in their assigned camp on one measure may fit just fine in most of the others. We are all mosaics.

And the bimodal nature of sex differences goes deeper than our outward anatomy. For example, there are a small handful of brain regions that show sex-specific differences, the best described of which is the *corpus callosum* (CC), a bundle of nerve fibers that connects the right and left sides of the brain and allows them to communicate. Research on sex differences in the CC goes back more than a century and the results have been all over the place, with many studies finding no sex differences whatsoever.[13] The advanced magnetic resonance imaging (MRI) techniques that matured in the early 1990s provided the "definitive" result that, on average (and only after dividing by overall brain size), women have a slightly thicker midsagittal corpus callosum (one of the four anatomical regions of the CC) than men do. So underwhelming is this difference that a computer algorithm trained on sex-specific CC measurements could correctly predict the sex of a brain only about 70 percent of the time. That's not that much better than a coin flip. And this is the brain area with the *best documented sex differences*.

Sex differences have been found in other brain areas as well, including the hippocampus, the amygdala, and the pre-optic area (the POA, discussed in chapter 2), and also in the proportion of gray matter (cell bodies) to white matter (nerve connections).[14] Some of these are equivocal, meaning the results have not been consistent across different studies. And others seem to depend entirely on the precise way that the measurements are taken, with different approaches giving wildly different results. Even still, hampering this entire discussion is the fact that the size and shape of different brain areas change over time, responding to how the different regions are used, a phenomenon called *anatomical plasticity*. Someone who spends hours every day on precise physical tasks will experience expansion of the relevant spatio-motor association areas, for example. So, in a world in which men and women, on average, still do different jobs and tasks, we fully expect that their brains will develop differences.

And remember, the brain differences that we see are "on average," and the ranges that we observe in males and females overlap considerably, just like height and weight. And further still, the measurement that an individual has in one brain area doesn't always sex-correlate with that of another brain area. A person may have a characteristically female corpus callosum but a characteristically male hippocampus, and so on. So, while we do observe a weakly bimodal spectrum of sex differences in the size of a few brain areas, if you consider an individual brain as a whole, even an accomplished neuroanatomist has no better than a 50 percent chance of predicting the sex using only anatomical measures. There is no such thing as a male brain or a female brain.[15]

Other parts of the body, from the size of the heart to fat deposits on the hips, show similar overlap in the male-female bimodal spectrum.[16] Some are more strongly sex divergent than others. Upper body strength is far more sexually dimorphic than lower body strength, for example. I have been a gym rat for years, but a close female friend of mine far surpassed my deadlift max within a few months of starting to work out. She now deadlifts a hundred pounds more than I do.

So when it comes to male and female bodies, what we end up with is a bunch of measures with bimodal distributions. But a bimodal distribution is not the same thing as binary distribution, for two key reasons. First, as already mentioned, each one of us ends up in different places on the spectra for different measures. We may have a typically male-shaped upper body but typically female-shaped hips and buttocks. Second, insistence on a binary ignores all of the data in between the two peaks. Just about the only measure in which the sex distribution is binary, rather than bimodal, is in the production of sperm and eggs, and that's why it is the final retreat of those committed to a sex binary. If we're going to ignore everything else, then this binary should be called *gametic sex*, not *biological sex*, since the only biology it considers is that of the gametes.

In my view, the most important reason to consider the entire body, rather than just the gonads, when we discuss biological sex is the existence of intersex individuals. Committing to the sex binary forces us to disregard all the people whose bodies deviate from typical expectations. Surely I can't be the only one who sees the inconsistency and cruelty of this approach. By making everything about sperm and eggs, we are elevating this one measure over all of the other aspects of male and female bodies, aspects that matter far more to us in every other context. Why?

Intersexionality

Around 1.7 percent of us are intersex.[17] This means that promoting the sex binary means marginalizing and stigmatizing *tens of millions of people* for no other reason than they frustrate our desire for neat categories. While it's true that the vast majority of individuals have either XX or XY chromosomes and their sexual anatomy develops accordingly, there are also variations on that theme. This topic is fraught because intersex individuals have suffered enormously through the ages, from cruel mockery and derision on the one hand to well-meaning but harmful medical intervention on the other. We are so committed to the sex binary that individuals outside this mold are seen as aberrant and met with horror. It's a tragedy that is totally unnecessary. To understand why, let's do a quick crash course in human embryology.

As embryos, humans have *bipotential gonads* that can develop as either ovaries or testes under the right signals. We also have two sets of ducts—one set that can develop into the typical male internal plumbing and one that can develop into the female internal plumbing.* Most of the time, either the male ducts develop and the female set disappears, or the female ducts develop and the male set disappears. And lastly, human embryos have folds of skin imme-

---

* The "male internal anatomy" refers to the seminal vesicles, bulbourethral glands, and vas deferens and develops from embryonic structures called the Wolffian ducts. The "female internal anatomy" refers to the uterus, cervix, fallopian tubes, and internal vagina (but not the vulva) and develops from the embryonic Mullerian ducts.

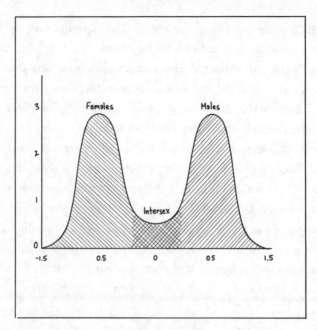

Bimodal distribution of sexual development

diately in front of the anus that can develop into either a penis and scrotum or a clitoris and vulva.* In summary, for the first eight weeks after conception, all humans have an unspecified gonad, both male and female internal plumbing, and androgynous skinfolds on the outside.

If an embryo has a Y chromosome, a gene on this chromosome initiates a process that turns the bipotential gonads into testes. The developing testes then begin to produce androgens, including testosterone, and another protein that triggers development of the male internal plumbing and blocks the development of the uterus and fallopian tubes. The androgens also "masculinize" the external anatomy, causing the folds of skin to develop into a scrotum and a penis.

If an embryo does not have a Y chromosome, female-typical development occurs by default at around ten weeks. The bipotential gonad will automatically develop into an ovary. With no testes, there are no androgens, so the male

---

* As an aside, the anus is the very first anatomical structure to take shape in a human embryo. We all start out life as nothing but an asshole. Some seem never to grow beyond that.

internal ducts wither away and the female ducts develop into the fallopian tubes, uterus, and the internal part of the vagina. Similarly, with no androgens, the folds of skin in the urogenital area naturally develop into labia, instead of a scrotum, and the apex of the folds becomes the clitoris instead of the penis. These are the most common developmental pathways for male and female fetuses through which they develop into infant boys and girls. Because this is the most common series of events, we call this *typical* sexual development, and about 98 percent of humans develop in one of these two paths.

But, as always, variations exist, and there are many ways that an individual can be intersex. In medicine, this variation is pathologized and referred to as *disorders of sexual development* (*DSD*), but the vast majority of intersex conditions are not in any way debilitating and most intersex individuals are fertile without medical help.[18] Therefore, it is purely a matter of perspective whether these conditions are disorders or simply anatomical or biochemical variations.

Historically, variation in sexual development was met with stigmatization, derision, discrimination, and worse.[19] Terms such as "hermaphrodite," "pseudohermaphrodite," and "eunuch" are among the more polite words used to describe those whose only sin was being born a little different. Worse, intersex individuals have been subjected to unspeakable atrocities, from abuse, abandonment, and euthanasia to misguided medical "corrections" that amount to mutilation and sterilization. Currently, the majority of intersex individuals lead perfectly normal lives with either minimal or no medical intervention. The intersex community is united in its opposition to such intervention until children can make their own medical decisions. We should listen to them rather than succumb to our fears about deviations from the norm.

The most common cause of intersex variation is called *adrenal hyperplasia*, a condition that causes the adrenal glands to secrete higher than normal levels of androgens. If this occurs *in utero*, the condition is called *congenital adrenal hyperplasia* (CAH). When genetically female embryos (XX) experience CAH, their external sexual anatomy can develop in the traditionally male pattern, with a scrotum and penis. However, this condition exists on a wide spectrum stretching from individuals with no perceptible deviation from the typical female presentation to those that appear 100 percent characteristically male, with most individuals somewhere in between. XX individuals with *late-onset adrenal hyperplasia* don't exhibit any deviation whatsoever until puberty, when the high levels of androgens begin to masculinize their bodies. In these individuals, their *chromosomal sex* (or *genetic sex*) may not match their *anatomical sex*.

Internally, genetic females with CAH will have ovaries because this is the default development when there is no Y chromosome to turn the gonads into testes. However, depending on exactly how high the androgen level is, the embryo may have typically female or typically male internal sexual anatomy, similar to the spectrum of feminized to masculinized external anatomy. Not surprisingly, genetic females (XX) have a more pronounced intersex experience with either type of CAH, whereas the bodies of genetic males (XY) are mostly unbothered by higher levels of androgens and no intersexuality results, although CAH causes other hormone perturbations that may have clinical consequences.

An intersex condition that is somewhat the opposite of CAH is *androgen insensitivity syndrome* (AIS), a genetic condition in which the cells of a genetically male individual do not respond to androgens in the typical way. When cells don't respond to a hormone, it is as though this hormone is not present, and since female-typical development is the default for both internal and external sexual anatomy, these individuals will be assigned female at birth and otherwise grow and develop normally as such. The first sign that anything is different is usually when puberty fails to arrive on time. The majority of XY individuals with AIS identify as female and take hormones to precipitate the feminization of their bodies, though in most cases they will not be able to have children. XX individuals cannot be afflicted with AIS, but they can be genetic carriers of it and pass it on to their XY children.

Some chromosomal differences can also lead to the intersex state. Due to incomplete separation of chromosomes during the formation of either sperm or egg cells, some individuals receive three sex chromosomes and are XXY, a condition called *Klinefelter syndrome*. Because humans have a Y-centered sex-determination system, these individuals develop as male in the typical way and almost always identify as such. Because our cells already have a system for using just one X chromosome or the other, as females do, boys and men with Klinefelter are often not even recognized until puberty, if at all. Differences are trivial and may include slight breast development, minimal body hair, and tall stature. Klinefelter men are often fertile but may need some help from reproductive technology.

The opposite of Klinefelter is Turner syndrome, a condition in which a female has a lone unpaired X chromosome instead of two, usually written as XO. Once again, because human cells are already built to handle either one or two X chromosomes, Turner syndrome does not cause much disturbance throughout the body. Turner females grow and develop normally, with a few minor differences such as short stature and slight webbing in the neck and feet. However, unlike Klinefelter men, Turner women usually need medical help to trigger puberty and to have children. Other than that, they live normal lives.

Failure to separate chromosomes can also lead to women having three X chromosomes, so-called *Triple X syndrome*. The vast majority of XXX women have no physical, physiological, or fertility differences, and it is estimated that at least 90 percent go totally undiagnosed. Similarly, some men have multiple Y chromosomes, and they are equally unaffected by this difference. There are even extreme cases of chromosomal nonseparation, leading to individuals with four or even five sex chromosomes in any combination. Generally speaking, females with multiple X chromosomes have few or no problems and genetic males (those with at least one Y chromosome) with multiple X chromosomes, no matter how many, are considered Klinefelter males. We are remarkably tolerant to differences in the number of sex chromosomes. That is, our *bodies* are tolerant of these differences. Our *culture*, not so much.

Collectively, CAH, AIS, and differences of chromosome number account for the majority of the intersex conditions that have a known cause. However, it is worth noting that many of the individuals with these conditions do not consider themselves intersex, either because their deviation from the typical male or female body is trivial, or because medical intervention is able to fully reverse whatever deviation they experience. There are literally dozens of other known causes of the intersex condition, from differences in hormone synthesis to subtle genetic variations in the sex chromosomes.

More important, for many intersex individuals, the precise cause of their deviation from typical sexual anatomy is totally unknown and appears to be sporadic. Embryonic development is astoundingly complicated, a tightly or-chestrated series of genetic and biochemical events that plays out in a tiny physical space and with very little margin for error. Anomalies, big and small, can occur throughout the body, and all of us have some of these quirks and flukes. Some have no genetic basis, but some are the result of subtle mutations in our genomes and can therefore be passed on. We are all born with a couple of hundred new mutations that we can contribute to the gene pool. Most of these are silent, but some cause slight tweaks in our anatomy, including our sexual anatomy.

The language we typically use for alterations in development is strange when you think about it. When an organ doesn't form in the typical way, we usually call this a "defect" or an error. But we also know that, over the course of evolutionary history, all anatomical innovations—anything new at all—begins as a mutation. From the placement of the blowholes on the backs of dolphins to the upright posture that changed our destiny as African apes, anatomical tweaks are the raw material for evolution's creative potential. They are *not* errors or defects. They are simply variations. The constant barrage of mutations appearing in a gene pool is nature's way of introducing variation, and it does so blindly and randomly. Then sexual reproduction spreads that variation through the population and generates diversity. Without a doubt, variation and diversity are *good* for a species—essential, in fact.

The simple biological truth that variation is *good* is often lost on us because modern society has trained us to be suspicious of differences. But this is a *psychological* tendency that we would do well to dismantle. It harkens back to the era when prehistoric humans lived in small bands that were often in competition with each other, fostering a sense of tribalism and prejudice that still vexes us today and underpins some of our greatest social problems. Once we recognize that anatomical variation is nature's tool kit of creative innovation and progress, we can change the way we think about our fellow humans with differences in their sexual anatomy.

Consider the diverse approaches to biological sex described in chapter 1. From the penis-fencing flatworms to the sex-switching clownfish, these interesting variants all began with a subtle tweak caused by a mutation. What could have been described as a defect or aberration was actually the beginning of something new and creative. In my view, this perspective should inform our approach toward people with intersex differences. They are not defective; they are merely different. As they mature, they may seek medical intervention or they may not, but let it be their own decision. Nature loves diversity. We should too.

In January of 1991, a girl named Mokgadi Caster Semenya was born in the tiny village of Ga-Masehlong, near Polokwane, South Africa. While in grammar school, she began competitive running and dominated every race she entered at the local level. Her domination continued at regional and provincial competitions, and she soon entered national and international youth competitions, culminating in the 2008 Commonwealth Youth Games, where she won the gold medal in the 800m race at the age of seventeen. Before her lay an exceptionally promising future in competitive mid-distance running.

In 2009, now an adult, Semenya competed in the world championships and won a gold medal by completing the 800m in 1:55.45, the fastest recorded time for that event anywhere in the world that year. Following this, she came under scrutiny by the International Association of Athletics Federation (IAAF, now called World Athletics) due to suspicions raised about the rapid improvements in her race times. Having shed 8 seconds from her best 800m time and a whopping 25 seconds from the 1500m race, she was suspected of using performance-enhancing drugs, commonly referred to as *doping*.

Apparently without her knowledge, the IAAF also subjected Semenya to biochemical sex testing. Although athletes are assured these tests are kept confidential, it was leaked to the press that she possesses a Y chromosome and was considered by the IAAF to be genetically male. However, there was never any doubt that she is anatomically female and was raised as such. With no evidence of doping, the IAAF allowed Semenya to keep her 2009 gold medal and cleared her to continue to compete in world events. In 2010, however, the IAAF subjected her to another round of testing and she was once again cleared to compete. In 2012, Semenya competed in the Olympics for the first time and finished second in the 800m. That silver medal was later upgraded to gold when the original winner was found guilty of doping and had all medals stripped from her.

Semenya again won the gold in the 2016 Olympics, which unexpectedly ignited a worldwide conversation about the role of testosterone in the performance of elite female athletes. In response to this controversy, the IAAF revised its policies in 2018 and set new standards for acceptable testosterone levels, a move widely seen as specifically targeting Semenya, since the new rule applied only to three events: the 400m, 800m, and 1500m races. She was forced to either withdraw or be subjected to medical interventions to artificially lower her circulating testosterone levels to within the acceptable standard. Semenya objected, having already taken these medications previously in an attempt to appease regulators and suffering side effects she described as unbearable. Legal challenges ensued, with Semenya mostly losing her appeals and the rules were allowed to take effect. Semenya has appealed all the way to the European Court of Human Rights, which ruled in her favor in 2023. However, their ruling served only to punish Switzerland, where World Athletics is based, does not reverse her banishment, and Switzerland's appeal is still pending. The summer of 2024 saw Semenya once again on the sidelines for the Olympic Games in Paris.

The case of Mokgadi Caster Semenya raises thorny questions about sex and gender. How can a female have a Y chromosome? Does a Y chromosome give a specific advantage in athletic performance? Does testosterone? How much testosterone is "too much" for a female to have? The questions get even more nuanced as the IAAF proceeded to not only set precise standards for testosterone levels but also carve out exceptions for various medical conditions (though not for Semenya). The situation, in short, is a mess.

The mess, however, is not the fault of Semenya. She is not a mess; she is a woman, a woman with an incredible athletic talent. Semenya also harbors some variation in her sexual development that puts her at odds with a world hell-bent on defining people according to the size and shape of their genitals and, increasingly, the concentrations of various hormones circulating in their blood. Yes, the situation is a mess, but it is a mess of the IAAF's own making. With all the medical experts at their disposal, they still managed to trip over their own misguided attempt to enforce fairness.

In reality, the IAAF chose to establish a baroque and constantly changing series of standards and exceptions that are incredibly *unfair* to athletes such as Semenya whose only offense is existing as their own unique selves. On some biological measures, Semenya is outside of a range that some people consider "normal." Okay fine, but let's keep in mind that just about *all of us* will have some biological measures, out of the thousands we could consider, that place us outside of the normal range. "Normal" just means that it fits within a majority of the population, which could be good, bad, or neutral.

The sex testing of Mokgadi Caster Semenya, one of the fastest mid-distance runners ever, is an appalling chapter in the history of sports. Not

236 · The Sexual Evolution

only is sex testing humiliating and dehumanizing per se, it was initially done *without her knowledge*. Sex testing of female athletes for competitive sports has an ugly history that began in earnest in the 1950s with the parading of naked athletes for anatomical inspection by panels of doctors and other officials. This was replaced by chromosomal testing in the late 1960s, through which female athletes with a Y chromosome were disqualified. Following the discovery of androgen insensitivity syndrome (described above), a condition in which XY individuals can develop as almost fully typical females, this method was abandoned near the end of the twentieth century. The current regime of sex testing involves examining the concentration of testosterone, and this appears to be the test that resulted in the disqualification of Semenya from international competition.

Because of the confidentiality involved, the medical situation with Semenya is not entirely clear, but based on the limited information available and public comments by various officials and Semenya herself, it seems that she is genetically XY and possesses some degree of androgen insensitivity. This means that her body does not respond to testosterone as strongly as most bodies do. Her external anatomy appears to be female, as she was assigned female at birth and continues to identify as such. However, likely due to her condition, her body makes more testosterone than females typically do. This is, at least in part, because her body doesn't "see" the testosterone very well, so her adrenal glands make more of it in an attempt to compensate.

Using elevated testosterone levels as justification for banning Semenya from competitive sports is scientifically unfounded. Here's why.

First, I would argue that every single Olympic-level athlete harbors some form of natural biological variation that places them outside the normal range and that gives them an advantage in their performance. By definition, Olympians are well outside of the normal range of human physical abilities, and I would bet dollars to donuts that all elite athletes harbor mutations or other measures that partially explain their incredible abilities. Simply put, elite athletes aren't "normal." That's what "elite" means.

Olympic swimmer Michael Phelps has a genetic difference that drastically reduces the production of lactic acid in his muscles when they are oxygen deprived. In some contexts, this would be detrimental, but during sudden bursts of muscle exertion, it reduces fatigue and quickens healing afterward. To my knowledge, no one has suggested he be stripped of his many medals. Instead, this difference is hailed as part of what makes him great.

The same is true of the cross-country skier Eero Mäntyranta, who had a mutation that increased his natural production of erythropoietin, a hormone that promotes the production of oxygen-carrying red blood cells, giving him more than a 25 percent boost.[20] Importantly, should any athlete administer erythropoietin to themselves exogenously, they would be banned for doping, just as an athlete who injects testosterone would be. Meanwhile, Mäntyranta's natural form of doping was never punished, while Semenya's is.

The whole concept of "talent" is based on natural inborn advantages.[21] Sure, that talent must be developed through hard work (and no one suggests that Semenya didn't work her ass off), but clearly, we're not all born with equal potential regarding our physical abilities. No matter how hard I work, I will never swim as fast as Phelps, ski as fast as Mäntyranta, nor run as fast as Semenya. They have the genes for that kind of excellence and, presumably, I don't. That's not unfair; it's life. Why are we singling out this difference, slightly higher levels of testosterone, for exclusion?

And second, there is, to date, no direct evidence that Semenya's elevated testosterone plays any role in her incredible athletic abilities. In fact, her androgen insensitivity argues against this. Her body doesn't respond well to testosterone. That's why the levels are high in the first place and why her anatomy developed in the female pattern. Even the authorities seem to recognize this as they chose to tailor the exclusionary rule to only the races that Semenya competes in, so as to avoid ensnaring other athletes in their misadventure. Punishing not only her particular kind of variation, but her specifically, reeks of the colonial indignance at Black excellence that has long characterized the way that European sports authorities treat athletes from the global south.[22] You may roll your eyes at that, but I hold a PhD in human physiology and I think it fits the evidence better than any scientific argument that's been offered.

Third, aside from a ban on the use of testosterone injections, there has never been an attempt to hobble the natural potential of male athletes with elevated testosterone. The stated goal in forcing Semenya to artificially lower her natural level of testosterone was to "level the playing field," so why do the authorities not take the same approach with male athletes? If a high level of testosterone constituted an unfair advantage for Semenya, wouldn't it also do so for male athletes, some of whose natural testosterone levels are off the charts? The logic here is inconsistent at best.

Mokgadi Caster Semenya is an incredibly gifted athlete who has labored under excruciating conditions to develop her talent and reach her full biological potential. She has taken no drugs or hormones to boost her performance beyond what she is naturally capable of. She has played by the rules, worked hard, and represented herself and her country beyond reproach. Her testosterone level is a little higher than that of most women, which may or may not contribute to her biological potential for running the 800m in under two minutes.

So why do we single her out for exclusion? The answer is simple and sad: because her variation is in the realm of sex. Our modern society has a rigidly enforced intolerance to that particular kind of variation. It does not have to be this way. In fact, it wasn't always this way. Variations in sex, gender, sexuality, and yes, even anatomy and physiology should be celebrated for the beautiful examples of diversity that they are. While the IAAF may feel she does not belong, Semenya fits perfectly on a planet filled with wondrous variety.

Gender Under Construction

In the final section of chapter 2, I described how the social dynamics of Virunga mountain gorillas had completely shifted over the past three decades, going from small single-male harems to large multi-male, multi-female groups with dozens of individuals living in relative harmony. This upended everything we thought we knew, not just about gorillas but about the flexibility of highly gendered behaviors and the stability of the social structures that spring from them. This raises pointed questions about the biological basis of our own gendered behaviors.

Central to the field known as *evolutionary psychology* is the notion that human behaviors are the product of natural selective forces that have, in the past, rewarded reproductive success. Though this claim may appear straightforward, many controversial corollaries have sprung from it. One that appears frequently in popular culture is the claim that some behavioral differences between men and women might be the result of how evolution shaped us differently, and these innate sex differences are the main reason some professions are dominated by men and others are dominated by women. For example, perhaps evolution shaped men to value professional ambition over a rewarding family, and it shaped women to value relationships over material things. It is true that we have observed small but statistically significant sex differences in the performance of certain cognitive tasks such as spatial and quantitative reasoning, and perhaps this is the reason we don't see more female astrophysicists on the Harvard faculty.

Setting aside the pesky fact that we have found more cognitive tasks in which women outperform men than vice versa, I probably don't need to shed much ink discussing the potential damage to gender equality these conversations can do. However, some will retort that if a given claim is supported by facts, it should not be silenced in deference to a social ideal. Steven Pinker, in particular, identifies strongly as a feminist and insists that doors of access must always be kept open. He would draw a line, however, if society pushes its social agenda so far that the potential achievement of some is sacrificed to develop the potential of others who may have less natural ability. Where that line might actually be found is difficult to say.

The problem, according to some, is that social engineering has costs. If men are, on average, naturally better than women at math, enforcing gender parity in top mathematics positions would mean elevating someone of lesser natural ability over someone with greater natural ability. The effect for society is a net loss in the fruits of our collective mathematics potential, quashed at the hands of social engineering.

But where do these supposed sex differences come from? Behavior begins in the brain and its associated neurochemistry, and differences in brain anatomy between the sexes are very few and very modest. More to the point, research has yet to exclude environmental causes for the few sex differences in cognitive abilities that we observe. No child on Earth has ever been reared free of the pervasive influence of socially constructed gender norms. As Elizabeth Spelke and many others have shown, even the most feminist parents begin treating their male and female children differently on their first day of life. Male babies are described as big, strong, and robust, while females are described as precious and delicate even when the actual size and strength of the infant does not warrant such distinctions. We are still very far from knowing what boys and girls might naturally aspire to and be capable of, absent the influence of a social environment that shapes them to be, and to want to be, different.

Evolutionary psychology, often the scientific bastion of reactionary thinking, may actually provide the answer to the very dilemma it raises. In his passionate attack on the "blank slate" view of the human mind promoted by Stephen J. Gould and many other progressives, Pinker convincingly establishes the genetic basis of certain behaviors as drives, instincts, and tendencies that provide a wide, but not infinite, range of possible outcomes. Underneath this is a principle as simple as it is elegant: genes plus environment equals behavior.

While blank slate proponents ignore the role of genes in creating behavior, evolutionary psychologists may be underappreciating the role of environment. In their worry about men whose wings are clipped to make room for women, they forget the girls whose preferences are shaped by a world that tells them how cute they are rather than how strong or how smart. In her book *Beyond Biofatalism*, Gillian Barker expresses the challenge saliently:

> Evolutionary psychologists focus on the Orwellian threat of overt unhappiness resulting from people's inability to fulfill their preferences, or from forcible attempts to suppress those preferences, but in so doing, they overlook . . . the danger of a hidden or cryptic loss of happiness resulting from a limitation of people's horizons that makes it impossible for them to form certain preferences in the first place.[23]

By obsessing over the costs and dangers of trying to overcorrect gender differences in professional achievement, evolutionary psychologists may be showing insensitivity to the very forces responsible for those differences. After all, their own defining principle emphasizes the importance of context in any consideration of how drives are turned into behaviors.

Like the mountain gorillas of Virunga, we too have accomplished wholesale behavioral transformations. Consider how different our behavior is compared to that of our prehistoric forebears described in the previous chapter, with whom we have negligible genetic differences. In fact, Pinker is one of the biggest champions of this. His book *Better Angels of Our Nature: Why*

Violence Has Declined chronicles the steady waning of violent deaths in human societies since the dawn of civilization.[24] Our constantly evolving social contract has allowed us to dramatically alter the way we interact with one another. Violence, even when warranted, is shunned in our youth and punished harshly in adults.

Pacification as a function of settled life is not limited to interpersonal violence; it extends to clan and state warfare. Just as laws and etiquette coerce individual behavior, treaties and sanctions nudge whole societies away from belligerence. Despite what sensationalist media headlines might have us believe, all forms of violent death have been on a steady decline for not hundreds but thousands of years. We are living in the most peaceful time in the history of our species, as least as measured by the odds of being killed or injured by violence.

What is this if not social engineering? The leviathan of societal and governmental supervision restrains our natural instincts toward competitive violence, and instead, we redirect those instincts toward sports, professional ambition, and the like. Genes plus environment equals behavior. The reason we don't hear any concern about the smothering of our "true nature" as violent apes is obvious: we are all better off for having our behavior shaped in this way. There is nothing "unnatural" about creating a social milieu that discourages violence any more than there is something unnatural about the way that Rwanda's mountain gorillas are now living. Instincts are vague and only take shape as behaviors within a specific environment.

There are other examples as well. In developed countries, we have drifted away from multigenerational communal living with grandparents and extended relatives all bunched in together. Instead, we much prefer a cozier domestic arrangement usually limited to one immediate family per domicile. What was once a pervasive social structure now seems totally unmanageable— unnatural even—to those of us who grew up knowing nothing but single-family households. We've seen a complete shift in just a few generations.

Many evolutionary psychologists also distress over possible reductions in personal freedom that could result from pursuit of a social ideal. However, loss of freedom is not always a bad thing. By prohibiting murder, we limit the success potential of a powerful person who might be otherwise inclined to murder his rivals and extend his success. But we still prohibit murder. By prohibiting theft, we limit the freedom of a worthy figure to seize resources from less powerful persons that he might be able to put to "better use." But we still prohibit theft. Matters of criminal justice are not the only areas in which we regularly apply social pressure to shape behavior. Public education, religion, role modeling, team-building exercises, even the Boy Scouts and Girl Scouts could be called forms of social engineering. Once again, a lion of evolutionary psychology, David Buss, has said it best: "Human behavior is enormously flexible—a flexibility afforded by the large number of context-dependent evolved psychological adaptations that can be activated, combined, and sequenced to produce variable adaptive human behavior."[25]

In my view, the stunning rapidity with which women have entered the highest levels of the workforce, compared to just a generation ago, argues forcefully that professional ambition in women is hardly unnatural. It seems much more likely that keeping women away from professional life was unnatural *in the first place.* In this light, social efforts to facilitate gender equality and parity may not be coercive social engineering but rather its opposite: the *removal* of coercion that has restrained women for far too long. It's a course correction. Just like a literal course correction, one has to expend some energy when altering the trajectory of a body in motion. A cruise ship doesn't corner quickly. In other words, there will be costs.

Questions of differing potential and preferences will likely persist for ages, but the observations of the mountain gorillas in Rwanda begs the question: If gorillas can transform their entire social structure, flipping highly gendered behaviors totally around—*and flourish in the process*—can we not also? The answer is simple. Of course we can, because we already have.

Chapter 8
The Gay Gene

While attending a science conference recently, I found myself backstage with a fellow speaker, one of the most recognizable scientists and public intellectuals in the world. There he was, holding court with admiring fans and pontificating about various topics, when the discussion turned to sex and gender. Given that I was writing this book at the time, I jumped at the opportunity to join the conversation and get his thoughts. We had an engaging and wide-raging conversation. He's clearly a progressive and would agree with much of what I say in this book. We sparred on a few topics, but in a friendly way.

After about an hour weaving in and out of some of the grandest questions about life and the future of our species, he asked me a question that stopped me cold. I suddenly realized just how common some major misunderstandings are regarding the biological basis of sexuality and attraction.

"You know, Nathan, one thing I wonder about is whether we will see homosexuality eventually start to disappear from the population now that it is more accepted and open."

I gave him an inquisitive look.

"Well, think about it, for thousands and thousands of years, homosexuality was forbidden and gay people had to live in the closet. Most of them just forced themselves into heterosexual marriages and had kids, so the genes for homosexuality persisted."

The look on my face went from inquisitive to confused.

"And now that gay people have the acceptance they always deserved, they don't have to live in the closet and they aren't forced into heterosexual marriages. That means they won't be breeding nearly as much. And so I wonder if homosexuality will just eventually disappear since the genes won't get passed on."

By now, the look on my face was full-bore incredulous, even agitated. His entire line of inquiry rests on deeply flawed assumptions. In that moment, it became crystal clear to me that misconceptions about even the basics of sexuality are widespread and deeply entrenched in our thinking. If a highly educated, progressive, and brilliant scientist could get so many basic things wrong about the biology of sexuality, how do the rest of us stand a chance?!

It's time to dispel some mythology about sexual orientation.

An Answer in Search of a Question

A great deal of research has been conducted on the specific question of what makes people gay. Because most of this work was framed within the dusty old debate of nature versus nurture, the question is usually construed: "Is homosexuality genetically programmed or is it the result of environmental influences?" To our modern understanding of sexuality, most of this research seems naïve and even problematic. Implicit in that research question is the assumption that heterosexuality is the normal, default state, and that homosexuality is an oddity that must have some specific cause. Sure, plenty of scientists do study sexual attraction and focus exclusively or mostly on heterosexuals, but no one ever asks what *causes* heterosexuality. That seems like an absurd question, right? Asking what causes homosexuality is just as strange.

In fact, I was conflicted about whether this book should include discussion of this kind of research because it could seem to indicate tacit approval of the notion that same-sex sexual attraction requires an explanation. The more we attempt to provide such an explanation, the more legitimate the notion becomes. In the many articles that have been published on the biological basis of homosexuality, there is a subtle but unmistakable subtext that the work was done in search of "what goes wrong." There must be a "gay gene" at work, or maybe gays have too much of the wrong hormone. But the study of animal behavior has shown that variation in drives and attractions are part of the normal sexual nature of any social species. As we saw throughout chapters 1 through 5, diversity of sexual behaviors is not a bug; it's a *feature*.

For more than 500 million years, sex has been an integral part of animal life. This was true even before animals evolved much of a social life, but things really exploded when sociality and cooperation emerged. At the end of chapter 3, I discussed some recent work in which biologists are finally beginning to question the notion that "deviant" sexual drives require an explanation; they are instead embracing the new paradigm that the animal sex drive is not narrowly purposed toward opposite-sex partners and procreation. Instead, bi- or pansexuality was the original sexual attraction, which was then tweaked as different animal groups emerged and evolved. Besides being simpler and therefore more likely, this scenario makes much more sense given all of the ways that sex can serve an animal's interests. A less restricted sex drive is more advantageous and more consistent with the evidence. Homosexuality is an innate feature of animal life and always has been.

Moreover, sexual attraction is clearly a complicated matter that will defy attempts at a simple explanation. Complex traits—especially psychosocial ones—could never be encoded by a single gene, or even just a few genes or other factors. As a general rule, the older and more complex a trait is, the more intricate its molecular basis. Relatively straightforward and "recent" traits like blood type, eye color, or lactose tolerance may be encoded by a single gene or a small number of genes, but older and more complex traits, like the shape of the brain or the way that our kidneys function, will involve contributions from hundreds or thousands of genes that have been evolving for millions of years. This is why you've never heard anyone speculate about the "brain gene" or the "walking gene." The sex drive—a feature of nearly all animals— originated hundreds of millions of years ago, long before animals evolved brains or kidneys. It is therefore likely to have been constructed through the action of many genes.

Is homosexuality genetic? Yes, of course, but it is not genetic in the way that having green eyes is genetic. It is genetic in the way that having kidneys is genetic. Our sexuality, like our kidneys, springs forth through the actions of thousands of genes acting in millions of cells as our brains and bodies take shape during embryonic development, through childhood and puberty, and even well into young adulthood. Furthermore, it seems clear—both in animals and humans—that sexual attraction is the result of both genetics *and* environment. For example, if one identical twin is gay, there is a 50 percent chance that the other is also. Clearly, genetics plays a big role, but it is not the only role. Why isn't the correlation 100 percent? We don't know. Among fraternal twins, the correlation is much lower, about 20 percent, but this is still far higher than we would expect by chance. Complicating this, most twins have both genetics *and* environment in common. Among identical twins that were separated at birth and raised in different homes, their sexualities correlate less than those raised in the same household, but not by much. And this is further complicated by the fact that even twins separated at birth experienced the same gestational environment for nine months, and there is strong evidence that sexuality takes shape, or is heavily influenced, during this time.

In addition, we cannot approach homosexuality as though it were a trait all by itself. It's a flavor of a trait that everyone has—the sex drive. There isn't a gene for green eyes; there are genes for iris pigmentation, and some versions of those genes lead to a green color. Even if there were a purely genetic basis for homosexuality, it would be due to the inheritance of specific versions of many important genes.

More to the point, everything we learned in chapters 1 through 5 about animal sexual behaviors strongly argues against a purely genetic basis for homosexuality. Instead, the emerging picture is that animals tend to have a vague sex drive with a range of possible manifestations that then takes shape based on interactions with the environment. When different drives or behaviors find success, they are selected; when they don't, they aren't. We can think of this, not as a blank slate, but as a range of possible outcomes. Nature may establish a range, and nurture helps shape the trait within that range. Genes plus environment equals behavior. While this may be difficult to imagine for an individual, if we think about a population, in which all the individuals have their own unique social experiences, it's easier to imagine diversity, experimentation, and differential success.

With this in mind, we can explore some of the research into the so-called biological basis of homosexuality. As misguided as the question may be, perhaps ironically, the lack of any strong conclusions from this research is informative in and of itself. There is a reason why we can't isolate this variable. Along the way, this work has led us to some interesting discoveries regarding how sexual attraction forms in human beings and other animals.

Lifting the Veil

Because of shame and taboos, sexual attraction and orientation did not become the subject of serious scientific research until the 1930s, beginning with the pioneering work of Alfred Kinsey.[1] Prior to that, the concept of *sexual orientation* did not exist as we think of it today, nor was there terminology to track it until recently. That's how strong the taboo on nonheterosexual sex was: we didn't even have words to discuss it. The word *homosexual*, a Greek-Latin hybrid originally constructed in German, only came into existence in the mid-1800s. Tellingly, the first known appearance of the term was in protest of an anti-sodomy law in Reichstag Germany.

For centuries, *sodomy* was the only term we had for homosexuality, but it was poorly defined, inconsistently used, and referred only to sex acts, not to homosexual attraction or orientation.* Sodomy most often referred to male-male sex and specifically anal sex, but occasionally also included heterosexual anal sex and even oral sex. Sometimes it included other male-male sexual contact, or else those acts could also fall more generally under the umbrella term of "perversion." Sodomy could sometimes include bestiality as well, and it came to mean any sexual crime "against nature." *Sodomite*, on the other hand, has always been used as a pejorative to refer to men who have sexual contact with other men (or want to). Similarly, sodomy is exclusively used in the context of legal prohibition or moral disapproval, never as a neutral descriptor of sex acts or gay people. So the only words that existed for discussing gay people were not just negative in their connotation, they were just about the most undesirable label that existed.

---

* Sodomy, of course, derives from the "sin of Sodom," the ancient Canaanite city that was destroyed by God in the Old Testament (and the Qur'an), along with Gomorrah, Admah, Zeboim, and Bela. Oddly, it is not at all clear from the text itself that same-sex sexual acts were the trigger of God's wrath toward Sodom. The Genesis account refers to a whole host of immoral behaviors as the reason for God's decision to destroy the cities, and the other books of the Old Testament, when referring to the wickedness of Sodom and Gomorrah, mention those other sins as well, sometimes without mention of the sexual transgressions. In fact, judging by the biblical texts themselves, arrogance and rudeness seem to be the relevant offenses.

Alfred Kinsey was the first to attempt a scientific study of human sexuality, and his work is considered foundational for the field of sexology. Before he turned his attention to human sexuality, Kinsey was an entomologist and he contributed millions of specimens to the insect collection at the American Museum of Natural History. Quite appropriately given the thesis of this book, Kinsey became interested in human sexual behaviors after observing the nonreproductive sexual behaviors of gall wasps. In fact, he developed the first version of the famous Kinsey scale for ranking orientation from homo- to heterosexual for use on wasps. It later occurred to him that what he was observing might also apply to human sexuality.

It is difficult to overstate just how poor the state of knowledge about sexuality was prior to Kinsey's groundbreaking work, and for this reason, not all of his conclusions and theories have stood the test of time. This is how science works. Research builds on, refines, and refutes work that came before it, but all of it contributes to the process of "honing in" on the true nature of things. The work done during the early years of a scientific discipline can later appear naïve, misguided, and outright incorrect. Not only that, as methodology and professional ethics develop, earlier incarnations appear amateurish and even unethical. The cumulative construction of knowledge is never a clean or linear path. It's a process of exploration, and not all avenues prove fruitful in the end, but the search for knowledge would not progress without a sloppy exploratory phase. So while Kinsey's work used controversial methods, he may have exaggerated some phenomena while overlooking others, and his theories seem overly rigid and incomplete to us now, they were a gigantic leap forward in their time and Kinsey was a courageous and imaginative scientist for proposing them.

In his first book, *Sexual Behavior in the Human Male*, and again in *Sexual Behavior in the Human Female*, Kinsey presented detailed scientific evidence that homosexuality was a real and enduring orientation, not just the rare and perverse result of psychological damage, as had previously been believed.[2, 3] In addition, it was Kinsey who first legitimized the idea that some individuals are sexually attracted to both men and women, the orientation he called bisexuality. His work was seminal in kickstarting the gay rights movement and the sexual liberation of women more broadly. For example, as laughable as it seems to us today, Kinsey had to arm himself with a great deal of original research in order to forcefully argue that most women actually *enjoyed* sexual activity. *Gasp!*

The Kinsey scale, which ranks individuals from 0 to 6—totally heterosexual to totally homosexual—was the first of its kind and gave birth to the notion that sexual orientation is a fixed personal quality, not an aberrant behavior or a fleeting curiosity. It was also the first acknowledgment that sexual attraction can exist on a spectrum rather than a strict binary, something that sexuality researchers have only recently begun to carefully study. He even later added a rating of X to those with no sexual appetite at all, an orientation we now call *asexual*. What was most shocking to 1950s America, however, were his claims that nearly half of all men had "reacted sexually" to another man at some point during their adult lives and that over a third had a full-fledged homosexual experience. The numbers for women were similar. As these stories hit the major newspapers, millions of Americans began to suspiciously wonder about the skeletons in each other's closets. Kinsey also reported that about 10 percent of adults were more or less exclusively homosexual and a similar number were bisexual.

Kinsey went further to describe the frequency of marital and extramarital sex among his study population and the prevalence of other "sexually deviant" behaviors including sadomasochism and even pedophilia. His books were bestsellers and made him a household name. The effect of his work was to awaken the Western world to the reality that sexuality was not as simple as we had believed. It is an awakening that continues today.

The work that Kinsey began in the 1940s sparked conversations and debates throughout society, from family dinner tables to academic research centers, and church pulpits to Hollywood awards shows. While the sexual upheaval of the 1960s was ostensibly focused on the uncoupling of sex from procreation for heterosexuals, the resulting relaxation of sexual mores more generally began to include homosexuality, and the gay rights movement officially began with the Stonewall riots in June of 1969.* The topic of sexual orientation, both as a cultural phenomenon and as the subject of academic research, was beginning to emerge from the cloud of taboo. In fact, through

---

* Upon the 2011 passage of marriage equality in New York, many of us spontaneously flocked to the Stonewall Inn to celebrate and pay homage to our forebears.

the 1970s and 1980s, arguably the most common rallying cry of the gay rights movement was "We're here; we're queer; get used to it," emphasizing that the taboo about discussion of sexual orientation was the first and biggest impediment to liberation.

We now know that deviations from strict heterosexuality are indeed common among human beings of all ages, races, ethnicities, cultures, nationalities, and religions, but the prevalence of queer sexualities is hard to know because results vary based on when, where, and how it is measured. The wording of the question can make a big difference, with numbers reported in various surveys ranging from 3 percent to 20 percent, with most falling between 5 percent and 10 percent.[4]

Personally, having pored through dozens of these population surveys, I think the most accurate way to measure queer people is to ask about strict heterosexuality and then subtract. Indeed, going about it this way gives much more consistent results from study to study, and I believe this is because it cuts through the variability in how queer people identify themselves as such. There are attractions, behaviors, and identities, and these don't always align, but asking someone if they are "attracted only to members of the opposite sex" seems to give a pretty consistent result. For example, the 2015 National Health Interview, a survey of over ten thousand Americans conducted by the Center for Disease Control, revealed that around 90 percent of men and 80 percent of women reported that they were attracted only to the opposite sex.[5] Yes, this includes some mostly straight people with a weaker attraction toward the same sex, but since attraction is on a spectrum, there is no clean place to draw a line. I am comfortable concluding that 80 percent of women and 90 percent of men are pretty straight, with 10 percent of men and 20 percent of women being something else.

Looking at the "something else" is when things start to get incredibly difficult to pin down. One big complication is that *having* certain attractions doesn't mean that one acts on them. For example, a man who has only ever had sex with women (and with great enthusiasm) but who occasionally feels attracted to certain men would almost certainly describe himself as heterosexual in a survey or interview. So what about his occasional attraction to men? How does that get counted? Further, being attracted to members of the same sex, and acting upon that attraction regularly, does not constitute an identity for many people. Plenty of men who regularly have sex with other men don't identify as gay. For many, being gay is not just about attraction or even behavior, but about identity. This is why public health researchers often study *MSM*—men who have sex with men—rather than studying "gay men." They are interested only in looking at those who engage in a certain behavior and don't want to get all tripped up with issues of identity, attraction, or labels.[6]

I know this well myself. I am attracted to men, the vast majority of my sexual partners have been men, and I identify as a gay man. Yet I do have consistent attractions to women and have had sexual encounters with women, including long-term relationships that were sexually satisfying (for both parties, if I do say so myself). I clearly have some level of bisexuality, so why don't I identify as bisexual? I'm not entirely certain. I just don't. Sitting in something of an opposite scenario is Svante Pääbo, the recent recipient of the Nobel Prize for his work sequencing the ancient genomes of Neanderthals (and thus making the work of my research lab possible). For decades he identified as gay, but when he met his now-wife and fell in love, he realized that he must be bisexual. They remain happily married after seventeen years. It's almost as if, at least for people like Professor Pääbo and myself, our attraction is more to individual people, rather than to a specific gender. Imagine!

If we count bisexuality as only those who identify as such, we would be left with the impression that bisexuality was rare indeed, around 1 percent of the population. But if you include those who identify as gay or straight but admit "more than a fleeting attraction" to the other sex as well, the prevalence shoots up to around 20 percent—larger than the entire LGBTQ community by most traditional measures.[7] In my view, this extreme reliance on the precise wording of the question proves the point that attempts to categorize sexuality and attraction within nice neat labels are futile.

The matter of bisexuality is so squishy that doubtful attitudes about its very existence are rampant, even in the queer community.[8] A common trope is that bisexuality for men is just the first step on the road toward coming out as gay. Others wonder if bisexuality is just the manifestation of an overdeveloped libido, a sexual appetite so strong that any partner will do. In 2020, in the largest and most comprehensive study of its kind, researchers confirmed that nearly all of the male research subjects who identified themselves as bisexual indeed demonstrate sexual arousal and attraction to images of both males and females more or less equally.[9] Although this doesn't seem particularly groundbreaking, press outlets around the world, from the BBC to the *New York Times*, reported on it. The newsworthiness of this research tells us all we need to know about public skepticism of male bisexuality. I imagine that millions of bisexuals the world over were relieved to discover that they exist.

In sum, after seven decades of research into sexual orientation, we have found that somewhere around 10 percent of the population is solidly queer, with another 20 percent or so that, while identifying as heterosexual, have a nonheterosexual attraction that is strong or frequent enough to admit it in surveys. So while most of the population is straight, sexual diversity is plentiful. Furthermore, the concept of sexuality has at least three major planks: attraction, behavior, and identity. Of course, these usually go hand in hand, but they misalign often enough to complicate research on sexuality. Different studies can generate conflicting results that aren't easy to reconcile because their methodology is different.

The way I see it, the very difficulty that we have in defining these categories is telling us that *the categories themselves don't work*. As Kinsey himself said, "The living world is a continuum in each and every one of its aspects."

The Band of Brothers

During the 1950s, when Alfred Kinsey first postulated that homosexuality could be part of a natural spectrum of sexual attraction, the behavioral sciences were consumed with the dichotomous question of nature versus nurture, and so this kind of either-or thinking also dominated research on homosexuality. The questions posed went like this: Were gay people born that way or did something happen during their lives that turned them gay? Was homosexuality a trait encoded by genes or the result of some environmental stimulus? Could anyone become gay given the right environment? Or are some people hard-wired to become gay no matter what? After six decades of solid research on this question, we don't have all the answers, but we know enough to say that—surprise, surprise—the dichotomy is a false choice. Sexuality is influenced by both genes *and* environment, by nature *and* nurture. We don't yet know if people are truly *born* gay, or, if not, for how long sexuality can be influenced by environment. But we do know that, for most people, one's orientation is well on its way by the time that puberty begins to kindle the fire of sexual attraction.

The first study that hinted at a biological basis for homosexuality was published in the 1960s by a very influential British psychiatrist named Eliot Slater, whose membership in the Eugenics Society was motivated by a strong belief that profound mental illness was primarily genetic and could eventually be bred out of the population. (He later claimed to have always been opposed to "coercive eugenics.") If that's not reason enough to be suspicious of his work with the homosexuals of London, he also maintained strong support for the use of the transorbital lobotomy—that is, the surgical tunneling through the top of the eye socket in order to suck out parts of the brain—well into the 1970s, long after the gruesome procedure had been abandoned by the medical establishment. None of this is particularly relevant to the groundbreaking paper that he published on homosexuality, but any discussion of his work should also include mention of how he thought about people who were different.

Over the course of five years, Slater examined the records of around four hundred men who had been hospitalized for sexual deviance and later "diagnosed" with homosexuality. In comparing their birth and family records, he noticed that their mothers were very often older than he expected, and when he looked closer, he realized why: the gay men tended to be younger siblings in larger families. This began as just an impression he had, so he needed a control group for comparison. The easiest group to get similar records for at the hospital were epileptics, so he examined the birth records of hundreds of men admitted for epilepsy over the same time period and compared the two groups. The statistics bore out his hunch: homosexual men tended to have more older siblings than epileptic men.[10]

This work seemed to implicate some kind of environmental influence on homosexuality, although no such birth order pattern was observed among lesbians. In the 1960s, average family size was still quite large in the West but was about to start shrinking, as the baby boomers prepared to leave their natal families and hormonal contraceptives became available. Within these large families was the common trope that youngest children, particularly boys, tended to be doted on by their mothers. After all, youngest children are able to hold on to motherly attention for a longer period than their older siblings because they aren't forced to give way to younger ones. In fact, mothers of this era often referred to their youngest as the "baby of the family" for years after they had grown up and, if they were boys, the term "mama's boy" was often attached. With this background, and with Freudian psychology still dominant, the obvious villain on whom the blame for homosexuality would land was mothers.

For about three decades, the folk wisdom on homosexuals was that they were the product of overprotective mothers. "Distant fathers" were sometimes also blamed, but mostly as a function of their failure to stop mothers from smothering their sons. I probably don't have to say too much about the harm that this kind of thinking did on all involved, especially young boys. The moral panic about homosexuality mixed a little too nicely with the general postwar patriarchy to create a soup of toxic masculinity: boys must be "toughened up," and gentleness and emotional expression should be minimized. To say nothing about how cruel and damaging this was, it was also ineffective. No amount of emotional and physical abuse would beat the gayness out of poor young boys, but the goals of toughening them up did work to the opposite of the intended effect: as the decades marched on, gays and lesbians marshaled their strength and independence to come out of the closet and begin fighting for recognition and equality.

Slater's observation that gay men tended to have older siblings was repeated in many different countries and was generally accepted as a real albeit mysterious phenomenon, and it received a powerful update in 1996: it turns out that gay men don't just tend to have more older siblings; they specifically tend to have older *brothers*.[11] The presence of older sisters doesn't seem to matter at all and, even more important, the effect does not hold if the older siblings are the result of adoption.[12] An increase in homosexuality is linked to having older *biological* brothers, regardless of whether the men in question were raised with those older brothers or by their birth mother at all. In other words, the trend was indeed because of nurture, but not in the way that psychologists had been thinking. If overprotective mothers don't make children gay, what does?

This phenomenon was officially named the *fraternal birth order effect* (FBOE) and is probably the strongest piece of evidence we have that biological factors influence sexuality in a specific way.[13] Given that it is seen even when the older brothers are stillborn or die in infancy and in younger brothers that are not raised with their older siblings, all signs point to the prenatal environment, rather than the childhood household, as the "causative agent."

However, it is also important to note how subtle the effect is. For a first male child, the odds of being exclusively homosexual (Kinsey-6) are about 2 percent. For every additional male child, the odds jump by about a third: a second male child has a 2.6 percent chance; a third, 3.5 percent. The fourth male child has a 4.6 percent chance of being gay, and if a mother has a fifth boy, he has a whopping 6 percent chance of being gay. This strikingly linear mathematical trend, which has been confirmed in dozens of studies, is convincing evidence that the effect is real and biological.

Additional corroborating but confusing evidence is found in other biological factors that appear linked to FBOE. For example, the number of older biological brothers (but not older sisters) also seems to lower birth weight for infant males (but not females), and those who will end up homosexual tend to have even lower birth weight on average than those who grow up to be heterosexual. So there appears to be a connection among birth order, birth weight, and the likelihood of being gay. Strange, but consistent.

Fascinatingly, the effect is also limited to men who are right-handed.[14] If we look only at left-handed or ambidextrous men, the FBOE disappears altogether! It's still a complete mystery as to what connects all of these factors. However, considering that both handedness and birth weight (obviously) are determined *in utero* and the FBOE is linked only to biological older brothers, not adoptive ones, it is clear that something in the fetal environment is different in a uterus that has previously carried male babies. Whatever that something is increases the chance that additional male babies will grow up to be gay.

One thing that might occur in a mother's womb after gestating male babies is that her immune system raises antibodies against male-specific factors.[15] Because women do not have a Y chromosome, their immune system views the proteins encoded by genes on the Y chromosome as foreign. Like all foreign proteins, male-specific proteins could be subject to immune attack by a pregnant mother, a phenomenon called an *alloimmune response*. Importantly, when we raise antibodies to a foreign protein, each subsequent exposure further enhances the antibody response, hence the use of "booster" vaccinations to maintain immunity. If a mother begins to make antibodies against male proteins in the fetuses she carries, the effect would be potentiated with each subsequent pregnancy. This could explain the FBOE itself *and* why the correlation gets stronger the more older brothers one has. This tentative hypothesis holds support among many researchers in this field, and the hunt is on for which male-specific protein(s) are at the center of this.

The first male-specific factor to be studied as a possible suspect in the alloimmune response in mothers of gay sons was the so-called *H-Y antigen*, a male-specific protein involved in spermatogenesis.[16] However, research into the maternal antibodies to this factor have yielded mostly negative or inconclusive results. In 2018, a new candidate emerged: a protein called *neuroligin*. This male-specific protein is a much more attractive candidate because it functions in the brain and is important in the development of the nervous system in male fetuses. The 2018 study found that mothers of gay sons had stronger immune responses to neuroligin than those with only heterosexual sons, and the effect was even stronger when the gay sons had older brothers.[17] These observations start to make the case that neuroligin contributes to the biological basis of homosexuality in men, but more research is needed.

Even if future research casts doubt on neuroligin, the case for the uterine environment somehow affecting the development of homosexuality in men is pretty strong, even though the effect is subtle. However, as with basically all psychosocial phenomena, relationships between cause and effect—or correlation and causality in technical terms—are seldom clear cut. Most men with older brothers are straight and plenty of gay men are firstborn sons or only children. Homosexuality appears to be determined as a probability, meaning that every human born has some small chance of being gay and some have a slightly elevated chance.

At the risk of falling victim to the anecdote effect, I cannot resist the temptation to close this section by mentioning that my husband and I both have two older brothers.

Fertile Myrtle

The fraternal birth order effect is not the only evidence for a biological basis for male homosexuality. Another was discovered at the National Institutes of Health in the 1990s and again in the early 2000s by an Italian research team. When considered along with the FBOE, this second biological connection has led to some interesting, if speculative, sociological possibilities.

In the Italian study, researchers recruited around two hundred men at random, half of which identified as strictly homosexual and the other half as strictly heterosexual. (Remember, scientists prefer hard categories, not fuzzy borders.) They collected family history and demographic information from the men and looked for differences between the two groups. They again noted that gay men tended to come from larger families, as predicted by the FBOE. However, they also observed something else quite unexpected. It wasn't just the mothers of the gay men who had lots of children. Their grandmothers and aunts tended to have larger families as well![18] This result is worth digging into a little bit.

In this study, the mothers of the gay men tended to have, on average, 2.69 children, while the mothers of the straight men had 2.32 children. That's not a huge difference, but it's statistically significant (at $p=0.02$ for the statisticians, meaning it has only a 2 percent chance of being due to a sampling error). However, that difference could be explained by the FBOE, so the researchers then compared only the firstborn males. In this comparison, the mothers of the gay men had 1.94 children, while the mothers of the straight men had 1.77, a smaller difference that was not quite statistically significant.

The real surprise came when the researchers looked beyond the first-degree relatives. The maternal grandmothers of gay men had 3.55 children on average (Italian families were larger in the previous generation) while the maternal grandmothers of straight men had 3.39 children. An even more significant difference was found when the researchers considered maternal aunts. Gay men's maternal aunts had 1.98 children while the maternal aunts of straight men had only 0.97 children. In summary, the mothers, maternal grandmothers, and maternal aunts of gay men all averaged more children than the same relatives of straight men. This implies some kind of connection between male homosexuality and female fecundity, at least in these Italian families.

Intriguingly, the effect was specific to maternal relatives. There was no difference in family size among the maternal uncles, nor the paternal aunts, uncles, or grandparents. It is not the case that *all* female relatives of gay men have larger families; it is specifically the *matrilineal* female relatives that are more fecund. Matrilineal relatives are those that are connected only through females. Your mother, siblings by the same mother, maternal grandparents, maternal aunts, uncles (not by marriage), and cousins (through the same maternal grandmother) are your matrilineal relatives. Your paternal grandparents, aunts, uncles, and cousins are not matrilineal relatives, since the connection between you and them cannot be drawn through only women.

This study has since been repeated and expanded upon, by the same research group and by others. Studies with thousands of participants have confirmed, in a variety of nationalities and ethnicities, that the matrilineal female relatives of gay men tend to have significantly more children than the same relatives of straight men, and that this difference is not seen in patrilineal relatives. Intriguingly, the effect is also seen in bisexual men and to the same degree as in homosexual men.[19] To nonbiologists, this connection to the matrilineal relatives of men seems utterly mysterious, but to a biologist the connection is clear: the X-chromosome.

As discussed in chapter 1, humans have Y-determined sex determination, meaning that genetic females have two X chromosomes, while males have an X and a Y. The Y chromosome has very little functional genetic information on it, but the X-chromosome has over two thousand protein- and RNA-encoding genes that have important functions throughout our bodies. While females get one X from each of their parents, males get their sole X from their mothers. Counterintuitive though it may seem, men get more genetic information from our mothers than we do from our fathers. While the puny Y chromosome passes from fathers to sons mostly unchanged, X chromosomes spread through both sexes and participate in the mixing and matching of gene variants we know as genetic recombination. The X chromosome is more like a regular body chromosome and harbors hundreds of genes essential to life. However, the unique way that the X chromosome is associated with sex determination means that it is inherited in a peculiar pattern within families. The genetic connection between gay men and their matrilineal female relatives may have a simple possible explanation: they have a good chance of sharing the same X chromosome.

The matrilinear connection between fecundity and homosexuality gave researchers a specific genetic target onto which they could focus their scrutiny, the X chromosome. But it also shifted the conversation regarding nature and nurture. The FBOE argues that nurture, specifically the prenatal environment, plays a role in male homosexuality, but the fecundity of matrilineal relatives argues that genetics—nature—is responsible. Instead of resolving the question, this new research further entrenched the nature and nurture camps, but, importantly, it gave further evidence that homosexuality was biologically, not psychologically, determined. Mothers were not to blame for creating gay sons, at least not in the way that sexists and Freudians had told us.

The connection between female fecundity and male homosexuality, apparently mediated by the X chromosome, led to some pretty hilarious speculation. As recently as the early 2000s, serious researchers were openly suggesting that there may be specific genes that are responsible for strong attraction to men. Some even suggested that these genes were so strong that they made men gay and females "hyper-heterosexual," and this is why female relatives of gay men were so fecund. Homosexuality had stubbornly persisted in the population despite the fact that gay men tend to leave fewer offspring because the uncontrollable horniness of these hyper-heterosexual women led them to have more babies, so the logic goes. Although some of the men that harbor these genes will grow up to be gay, any loss of fitness due to male homosexuality is offset by the higher birth rate overall.

Some speculators went even further and thoughtfully reminded us that gay men tend to be more generous toward their nieces and nephews than straight uncles are. There are indeed a few studies that have found that gay uncles spend more time and money on their young nieces and nephews than do straight uncles. (As if this is a genetic trait!) The idea offered by this crude pop evolutionary psychology is that some gene variants on the X chromosome make men gay and women overly sexual. Then those gay uncles are handy for helping to care for all those extra babies that the fecund women have.

I don't know about you, but this all sounds a little too neat to me. Sitcom stereotypes aside, the reality is that same-sex sexuality doesn't tend to reduce rates of reproduction, even if gay men tend not to have as many biological offspring in this particular cultural moment. Thankfully, the death knell for this cutesy theory would come quickly. But despite that, the perplexing connection between male homosexuality and female fecundity has been confirmed many times in populations throughout the world. The evidence for a weak but persistent genetic influence in male homosexuality is solid.

X Marks the Spot

Research in the 1990s and early 2000s seemed to confirm that gayness in men does come from the mother—not by overprotective parenting, but either through her X chromosome or antibodies that she raised during previous pregnancies, perhaps both. While research on the antibody theory has only recently begun, the genetic research has a longer history. Back in 1993, researchers at the National Institutes of Health studied gay men at an HIV outpatient clinic and, after confirming the FBOE yet again, they conducted what is called a *pedigree linkage study*. This kind of research attempts to correlate specific gene variants with some physical or behavioral feature within a family tree, in this case homosexuality. What they found was that the gay brothers whom they examined tended to share a handful of specific genetic variants with each other more often than they did with their heterosexual brothers and (as was discovered in a later study) more often than heterosexual brothers did with each other.[20] In sum, the researchers had found a cluster of specific genetic variants that preferentially appear in gay men, rather than straight men, within specific families.

It is important to note that the variants that are studied in this kind of analysis are not functional genes. They are more like signposts along the road. They are easy to track, but they don't usually have any purpose genetically.* So while this research cannot suggest a possible mechanism for the genetics of homosexuality, it can point to a specific spot in the genome that researchers can then examine more closely for clues. The region that the linkage analysis pointed to is called Xq28, a small gene-dense region at the very tip of the X chromosome.[21] This is yet more evidence for a maternal connection, and X-chromosome transmission, of homosexuality in men. Because this result

---

* The human genome is so full of useless material that I dedicated an entire chapter of *Human Errors* to so-called "junk DNA."

gave physical validation to the matrilineal inheritance of homosexuality, and because the studies combined unbiased molecular analyses with rigorous statistics, both the scientific community and the general public were convinced there was a genetic connection to homosexuality. We were tantalizingly close to finding the "gay gene."

In the region known as Xq28 there are hundreds of genes and dozens of them have been proposed, and later debunked, as the gay gene. The idea that a single gene would be responsible for homosexuality is absurd anyway: single-gene traits are easy to track in family trees and homosexuality does not show a simple inheritance pattern. But it is not unreasonable that a single gene could be a major contributing factor, among other factors, and so several research groups explored the role of Xq28 in homosexuality. Long story short: the results have been mixed. All of the follow-up studies conducted by the same research group confirmed a role for Xq28, while the results from larger studies by other groups have been less encouraging.[22] While Xq28 hasn't lived up to the original hype, most researchers in the field would have agreed (until recently) that the studies that have implicated it were well designed and that something is likely going on with Xq28 and male homosexuality.

Then came the fall of 2019. An enormous study involving almost half a million people, both men *and* women, examined many thousands of genetic variants with the help of the UK Biobank and the genetic testing company 23andMe. Previous studies had involved mere hundreds of participants, making this study far more comprehensive and inclusive (though heavily tilted toward Europe and North America) and providing far greater statistical power than any previous studies by far. This time, researchers asked the participants detailed questions about sexual activity, sexual fantasies, and sexual attraction, which also gives the study much more nuance. This was a truly massive effort.

The results? The study found literally hundreds of genetic variants spread all over the genome that preferentially associated with homosexuality.[23] However, all these variants that collectively contribute to homosexuality still account for only somewhere between 10 percent and 25 percent of its prevalence. In other words, this study argues that the collective genetic influence on homosexuality may contribute only up to a quarter, probably less, of the total prevalence of the trait. The remaining influence is likely a combination of environmental, epigenetic, random, and psychosocial factors.

In hindsight, the very notion of a single gene, or even a small number of genes, being responsible for homosexuality seems comically naïve. If there is one thing the genomics era has taught us, it's that all biological phenomena are far more complicated than we originally thought. Sexual attraction is a complex trait, far more complex than physical features like eye color or skin tone. So why should we expect the *genetics* of this trait to be simple?

Consider something as straightforward as height. Researchers have found over 800 genes that contribute to human height. *Eight hundred!* And what's more, collectively, these 800 genes constitute only 25 percent of the genetic contribution to height, meaning that another 2,400 genes play some role. Furthermore, all of these thousands of genes—*all of them*—do plenty more in our bodies than just influence how tall we are. The bottom line is that there is no one gene for height. Rather, thousands of genes influence height in some way while also influencing myriad other things, and nutrition and other environmental effects are important as well. And this is a physical trait that is way simpler than a neurobehavioral trait like homosexuality.

Of the thousands of gene variants that the researchers found to be associated with homosexuality, those located on Xq28 were not enriched among them, even when they focused solely on men. We can now say with as much certainty as scientifically possible that there is no gay gene.

You Shoulda Put a Ring on It

When it comes to the biological basis of homosexuality, a lot more research has been focused on gay men than on lesbians. There are several understandable reasons for this in addition to the predictable sexist ones. To this day, women are often left out of many medical studies and the male form is generally considered the biological standard, with women as a confusing variant all jumbled up by those crazy hormones. (Hormones that men also have, by the way.) But it turns out that there are some real differences that are partially to blame for this disparity, which scientists—a profession still dominated by men—have been all too eager to exploit as a reason to be satisfied with research that focused solely on men.

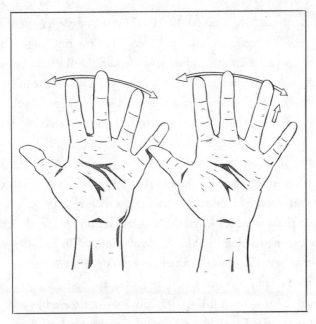

A high (L) and a low (R) 2D:4D ratio

First, in study after study, women do appear to be more flexible in their sexual attraction than men are, on average. Most men fit into the nice, neat categories that empirical science prefers and most women do not. Women self-report as bisexual, pansexual, heteroflexible, and other fuzzy categories more often than men do (although this may be changing). Whether this is the result of biology or a more permissive culture is still an open question, but I suspect that it is both. And second, the fraternal birth order effect and its connection to female fecundity are both phenomena that were waiting to be found and simply jumped out of the data when researchers began looking at the families of gay men. Neither of these phenomena are observed in lesbians, nor has anything else peculiar been observed in their family trees after a great deal of looking. We can't fault scientists for not studying an effect that simply isn't there.

But there *is* a biological phenomenon that has been observed among lesbians that seems even stranger than those we've observed in gay men: lesbians have shorter index fingers than straight women do! The actual measurement we're talking about here is called the 2D:4D ratio, or the length of the index finger (the second digit) divided by the length of the ring finger (the fourth digit). Lesbians, on average, tend to have a lower 2D:4D ratio, meaning that their index fingers are shorter than their ring fingers, while the index and ring fingers of heterosexual women tend to be roughly the same length. How this bizarre difference may have come about is a long and interesting story.

There have been spurious scientific musings about differing digit ratios between men and women dating all the way back to the 1800s. The first large-scale empirical study was published in the 1930s and reported that the preponderance of men had shorter index fingers than ring fingers, while most women had them at equal length. That observation sat mostly ignored for decades but came roaring back in the 1960s and 1970s as biologists began to obsess over sex differences in the human body. Recent studies have confirmed this subtle but consistent difference. The average 2D:4D ratio for men is ~0.95, while for women it is ~0.97. This may not seem like a large difference, but the standard deviation is low and the statistical significance is high. In a typical group of men and women, this difference can usually be spotted by just eyeballing each other's fingers. Interestingly, different ethnic groups also vary in their 2D:4D ratios, and in fact, those differences are larger than the sex difference.[24] However, the sex difference is observed within all ethnic groups examined to date.

The observation of a connection between 2D:4D ratio and sexuality came about in a roundabout and sexist way. In 1983, Dr. Glen Wilson, another psychiatrist from London, reported that women who described themselves as "assertive" and "competitive" had lower 2D:4D ratios.[25] Per his stated rationale, Dr. Wilson believed that assertiveness and competitiveness were masculine traits, so he hypothesized that women that were so incredibly masculine that they would call themselves assertive (gasp!) might also have some masculine physical features and he decided to examine 2D:4D ratio. If we look past the sexist notion that assertiveness and competitiveness are somehow unfeminine, it appears that Wilson "Mr. Magoo'd" his way into an interesting scientific result and even suggested a possible mechanism connecting the two traits: fetal exposure to sex hormones. His suspicion that the prenatal environment was involved stemmed from the fact that the 2D:4D ratio is established before birth and remains stable as our fingers grow to adult size.

Given the assumptions about assertiveness and masculinity, the logical hormone to explore for a role in the fetal environment was testosterone and, fortunately, there is an easy and noninvasive way to test the hypothesis that testosterone might play a role in 2D:4D ratio. In a condition called congenital adrenal hyperplasia, or CAH, discussed in the previous chapter, fetuses experience much higher levels of testosterone than usual. High levels of testosterone don't seem to have a big effect on the bodies of chromosomally male (XY) fetuses, but in females, it can be enough to drive their anatomical development in the male direction, depending on the timing and dosage of the testosterone, which is usually scant in a developing female fetus. In 2002, when scientists studied the 2D:4D ratio in people with various degrees of CAH, they got the confirmation for which they were looking. Both men and CAH-afflicted women show a more masculinized ratio (meaning shorter index fingers), and the effect is roughly proportional to the severity of the CAH syndrome that they exhibit.[26] In androgen insensitivity syndrome, which we can think of as the opposite condition as CAH, XY individuals develop as females and exhibit the typical female 2D:4D ratio.[27] Both of these conditions strongly hint that prenatal testosterone exposure is the link between sex and 2D:4D ratio.

Research in mice has definitively confirmed that androgens—the family of hormones to which testosterone belongs—are responsible for establishing the 2D:4D ratio.[28] During a precise window of time during embryonic development, androgens act to slow the growth of the first digit because it expresses more of the androgen receptor than the other digits (for reasons unknown). Therefore, the 2D:4D ratio—in mice as in humans—is a surprisingly good proxy indicator of the level of androgens that each of us were exposed to prenatally, and this ends up being connected to a wide variety of traits and clinical conditions.[29]

For example, in men, low 2D:4D ratio has been linked to incidence of prostate cancer, ADHD, aggressiveness, depression and anxiety, reduced empathy, conduct disorder, and penis size. In women, high 2D:4D ratio correlates with breast cancer, visual memory, anorexia and bulimia, and psychopathy. In both sexes, the 2D:4D ratio correlates with fertility and fecundity, numeracy, athletic performance, and verbal fluency; but in men it is a lower ratio that correlates with those traits, while a higher ratio predicts those traits in women. Hundreds of studies link the 2D:4D ratio to dozens of other traits, usually with very weak associations and always with a direct connection to known or assumed androgen effects.[30]

And now back to the lesbians. In 2000, a bombshell paper was published in the journal *Nature* that reported the results of a study comparing 2D:4D ratios from a large population of men and women, both gay and straight.[31] The study found, yet again, a small but statistically significant difference in the ratio between men and women; but when they broke the data down by sexual orientation, the women that identified as homosexual, on average, exhibited the smaller 2D:4D ratio that is more typical of men. For the first time, scientists had hard evidence of a biological connection to homosexuality in women to accompany the FBOE and maternal fecundity connection in gay men.*

A couple years later, the same research group dove further into the digit ratios of lesbians and asked if there was any difference between lesbians that self-identify as "butch" and those that self-identify as "femme."[32] First, we must admit that research in this area is fraught with controversy. The terms *butch* and *femme* are offensive to some people and many consider them cultural constructs more to do with gender expression than sexuality. Within the lesbian community, there is understandable mistrust regarding those who would do research in a way that seeks to divide the community into different "kinds" of lesbians. That said, the researchers approached the question by allowing volunteers to self-identify if they wish and they simply compared the finger length ratios of those who did.

The results were clear. Lesbians that identified as "butch" had low 2D:4D ratios, so low in fact, that, collectively, they were indistinguishable from men. So-called "femme" lesbians had high 2D:4D ratios and, collectively, were indistinguishable from straight women. Therein lies the controversy. Predictably, some interpreted this difference to mean that only the butch women were "real" lesbians, while more feminine lesbians were really just dabbling or heteroflexible or perhaps bisexual. Nonetheless, the lesbians themselves self-identified this way and the digit ratio difference was statistically significant. As my PhD adviser used to always say, "The data's the data."

---

* That first study also found a 2D:4D correlation in gay men, specifically those with older brothers, but that result has not always held up in other studies, while the result for lesbians has been more consistent.

Personally, I don't find it hard to imagine that, among gay men and women, there could be a spectrum of gender expression and a spectrum of sexual attraction, and that the two spectrums might overlap, if imperfectly. And I further find it plausible that exposure to androgens *in utero* could have some influence—and no one is suggesting that it is the *sole* influence—on both gender expression and sexual orientation. That is what the 2D:4D ratio is pointing to.

I am *not* suggesting that gender and sexuality are the same thing or even that they are inextricably linked, just that they have some common influences and that their expression *can be* linked in some ways. In lesbians, this may play out along the butch-femme spectrum, and I believe there is a similar spectrum among gay men. For example, some gay men identify as "masc" or "fem," whose gender expression are more traditionally masculine and feminine, respectively. But both are attracted to men. I don't find it offensive or problematic whatsoever to imagine that things like prenatal exposure to androgens or antibodies can influence where we all end up on that spectrum.

As long as we're all free to be ourselves and explore our gender identity and sexual orientation free from judgment, harassment, and violence, what difference does it make that there may be molecular mechanisms that helped determine that identity and that orientation? I would argue that it can hardly be any other way.

In sum, research on finger-length ratios in humans has found a connection among biological sex, sexual orientation, and fetal androgen exposure, providing further evidence that molecular events taking place *in utero* have some influence on sex, gender, and sexuality.[33] Importantly, this does not resolve the nature versus nurture debate because the uterine environment is essentially a combination of genetics and environmental exposures. If anything, it argues that the debate was misguided all along.

Baby I Was Born This Way

The work described above does not stand alone in the field of sexuality research. For example, there have been studies linking adult levels of the sex hormones—androgens, estrogens, and progesterone—to sexuality as well as gender expression, but the results have been weak, equivocal, or difficult to replicate. At one time, psychiatrists thought they could cure gay men by giving them testosterone injections. Not only did this not change the men's sexual orientation, it usually boosted their gay libido. Exposure to sex hormones *in utero* may influence sexual orientation, but later in life it does not appear to. Research on sexual behavior in animals has yielded similarly confusing and mixed results. Even with all the intense research on homosexual rams (chapter 4), we still don't have a concrete understanding of how sexuality takes shape in that species, let alone in any other.

What does this tell us? It could be that we just haven't stumbled upon the "right answer" yet. Undoubtedly, this is what researchers will say in their requests for more research funding. Given the results of the genome-wide study described above, I think we can safely rule out a purely genetic mechanism. However, an epigenetic mechanism—the effect of environment and past experience on how genes are expressed—could still be at play and, I suppose, is worth investigating. Similarly, while measurements of hormones and gonad size have yielded little in the way of consistent trends, other kinds of exposures and environmental stimuli, both sexual and nonsexual, could affect *how* those hormones affect our sexual appetites. As my graduate school endocrinology professor frequently reminded us, "When it comes to how hormones work, there are just a handful of different keys, but there are thousands of different locks and each opens a totally different door." We *know* that hormones play a direct role in sexual attraction because both appear coincident during puberty and deficiencies in one are usually accompanied by deficiencies of the other. But we have basically no clue about why or how.

Regardless, the mere fact that a simple answer has been elusive is itself revealing. It tells us that sexual attraction is complex, multifaceted, polygenic (controlled by many genes), and has been a part of animal physiology for a long, long time. For a good contrast, consider the diversity we see in human skin tones. There are just a few genes that account for all or nearly all of that diversity because skin tone differences are new to our species, having evolved in the last thirty thousand years or less. Sexual diversity, on the other hand, dates back to the very origins of sexual dimorphism in animals, *hundreds of millions* of years ago. That diversity is embedded throughout our constitution and can't be located in a few genes here and there. What's more is that, increasingly, we are observing that sexual attraction and orientation may be fluid anyway and they take shape based on the cultural milieu in which we are raised. That is surely part of why we now find ourselves in this time of sexual upheaval: the current culture allows for it. Our genes, hormones, and brains function in constant dialogue with each other and with the world around us.

We are who we are not only because of what we've inherited, but because of what we've experienced.

And now we can return to the misconceptions held by the famous-but-nameless scientist that I opened this chapter mentioning. To recap the question, he asked whether homosexuality might slowly be disappearing from the human population now that queer people don't have to force themselves into marriages and having kids, so the genes for homosexuality won't be propagated. Let's recount and dissect each misconception one-by-one.

Misconception #1—*Homosexuality is genetic.* Nope. This one is the quickest to dismantle. Literally hundreds of research studies, including large-scale, genome-wide association studies, have been conducted in an effort to discover the genetic elements responsible for sexual orientation, and the results have been underwhelming to say the least. While various individual projects have produced candidate genes and chromosomal regions as possibly involved, those results almost never hold up in subsequent studies. The results are all over the place, and the largest and most "definitive" of such studies, as explained above, identified literally hundreds of possible genetic elements, spread through every single chromosome, as having some correlation with homosexuality. Even worse (for the viability of this misconception), the collective contribution of all of those hundreds of gene variants accounts for no more than 25 percent of the incidence of homosexuality (and possibly far less).

So the only conclusion to draw is that either homosexuality is almost entirely nongenetic or that it is so diffusely genetic that it could never be "bred out" of the population. In hindsight, we should have assumed this all along because homosexuality has never been observed to run in families like other genetic traits do. And our friendly but mistaken scientist is correct that any genetic trait that leads its bearers to breed less should disappear from the population. And yet, sexuality that includes some degree of same-sex attraction is prevalent throughout all human populations and indeed all animal populations whose sexuality we have bothered to examine. If homosexuality is an impediment to reproduction, why is it so widespread in both humans and other animals? And that leads me to the next misconception.

Misconception #2—*Homosexuality harms biological fitness*. Nope. In biology terms, *fitness* is the quality of leaving many successful offspring. And so, to our modern sensibilities, this one seems to make a little bit of sense. After all, queer people are less likely to have children than heterosexuals, so same-sex attraction would seem to lower biological fitness. In fact, I sometimes joke to students and other evolutionary biologists that, as an adoptive father, I have *negative* biological fitness, since not only do I not have any biological offspring, I am investing all my energy and resources into *someone else's* genetic progeny.

Once again, this misses the point as well. Yes, in this one very particular moment in time, same-sex attraction may, on average, reduce the likelihood of someone leaving genetic offspring. But we have no reason to believe that these conditions will persist very long and every reason to believe that this is a temporary anomaly.

First of all, sex and sexual attraction are not very directly tied to the production of offspring. Remember all those nonprocreative functions of sex discussed in chapter 5? Animals, including humans, have sex with each other for a whole variety of reasons, and those various purposes do not compete with one another. Neither the same-sex pairings of albatross and penguins, nor the traumatic male-male insemination of bed bugs, nor the rampant sexuality of lesbian bonobos, nor the sexual congress of alpha and beta lions impede their ability to leave offspring. In fact, in all of those cases, the same-sex sexual behavior *enhances* biological fitness. Minority sexual orientations sometimes bring their unique advantages simply by virtue of being different from the crowd.

Second, even in today's climate, many queers are having children! Some lesbians and gay men have children from a previous heterosexual marriage or relationship; many opt to have children through artificial insemination or *in vitro* fertilization; gay men are frequent sperm donors, both for friends and acquaintances and through anonymous donation in sperm banks; and more creative and progressive family structures are appearing in which queer people are indeed procreating. As birth rates among traditional heterosexual families continue to plummet, and reproduction among sexual minorities continues

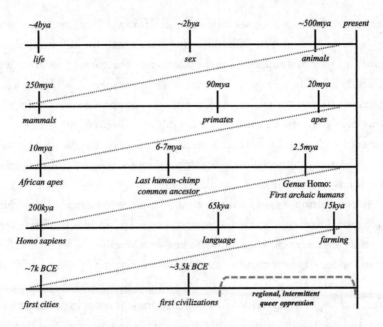

Nested timelines representing the origin of key taxa and phenomena in evolutionary history. (Bya, billions of years ago; mya, millions of years ago; kya, thousands of years ago; BCE, before common era). The summary of this is that oppression of sexual minorities has occurred in the tiniest fraction of our natural history.

to grow, we are approaching parity. And this has happened *within a single generation* of the lifting of oppressive strictures on homosexuality.

Misconception #3—*Homosexuality was oppressed for most of human history.* Nope. As we saw in the previous chapter, this assumption is also way off. There is about five thousand years of recorded human history, and even in the most repressive cultures and regions of the world, homosexuality was only specifically subjugated for a portion of that time, certainly not most of it. And in most parts of the world, it wasn't subjugated *at all* until Eastern and Western imperialism rained down its fire and brimstone. The restrictive and oppressive cultural prohibitions against nonheterosexual and nonprocreative sex that have dominated the last few centuries are an aberration within the larger history and diversity of various human cultures.

But even that understates the reality. Our species has existed in its present form for about two hundred thousand years and archaic humans such as *Homo erectus* have been walking the Earth for over 2 *million* years. Apes have been around for over 20 million years, mammals for 200 million, and animals for over 500 million. The deepest roots of sexual reproduction were laid nearly 2 *billion* years ago. The best evidence we have tells us that same-sex sexual behaviors existed, free from any form of oppression or restriction, during nearly all of that time. A few centuries of oppression, and now the reversal of such, is unlikely to unravel the effects of hundreds of millions of years of evolution, especially when you consider the above point that sexual attraction is not genetic or is very diffusely so.

Misconception #4—*Heterosexuals are purely heterosexual.* Throughout this chapter, I have concentrated scrutiny on sexual minorities and left heterosexuals largely unexamined. But the point I want to raise now is that we have a lot of reason to believe that so-called heterosexuals probably have a lot more flexibility and diversity in their sexual attractions than they realize or care to admit. Although the majority of the human population would describe themselves, and maybe even truly identify as, staunchly heterosexual, that does not necessarily mean that they were "born this way" and destined only for that specific sexual attraction.

To be sure, I am not saying that most straight people are closeted gays or bisexuals, nor that they are repressed and lying to themselves about their sexuality. What I am saying is that our biology doesn't hard-wire our sexual attraction in a super specific way. Instead, sexual drives and attractions are vague and take shape with input from our personal and cultural experiences. We know this from a variety of evidence including historical trends and by examining identical twins separated at birth and raised in different environments. Their respective sexualities can be quite different, and those differences can often be explained through an examination of their respective formative environments. An environment in which deviance from heterosexuality is brutally suppressed places a strong negative input on deviant attractions. Some individuals may develop same-sex attraction anyway, but many won't. This does not mean that we are all born as blank slates, but rather with something like a range of possible outcomes. Sexual "orientation" seems to be the result of how our biology interacts with our psychosocial environment during a crucial period of childhood and adolescence.

In sum, based on everything we know and for a variety of reasons, we should not expect that same-sex attraction and sexual behaviors will disappear even if queer people are currently not having biological children at the same rates that self-identified heterosexuals are. In fact, we should probably not expect homosexuality to disappear even if queers were to leave *no* children at all for thousands of years. Because there is some biological and genetic influence, it is possible that homosexuality would slowly wane if the negative selective force was strong enough and was applied for long enough. But if the centuries of queer oppression in Western and Eastern civilizations have taught us anything, it's that diverse sexualities are an integral feature of our species and can't be easily stamped out. And, given that the same is true for basically all other species, it probably couldn't be any other way.

The Queen James Bible

Those interested in English or Protestant history are quite familiar with King James VI—known in England as King James I—the monarch who ordered a new English translation of the Bible, which is still in widespread use, and which bears his name. The son of Mary, Queen of Scots, James became the king of Scotland at a very young age. When his cousin and fellow descendant of Henry Tudor, Queen Elizabeth I, died, he inherited the thrones of England and Ireland as well. This represented the full and final unification of those crowns, thus creating the monarchy of Great Britain and (Northern) Ireland as it still exists today, currently occupied by his great-great-great-great-great-great-great-great-great-great-great-great-great-great-grandson, King Charles III. What is not widely known about King James, the first monarch of unified Britain, is that he was a gay man and rather openly so.[34]

James met his first lover, and by all accounts the great love of his life, when he was just thirteen. Already the king of Scotland, the young teenager became smitten with Esmé Stewart, a French nobleman. The king barely hid his affections for Stewart, embracing and kissing him in public and sharing his bed with him frequently. He named him a "gentleman of the bedchamber," then appointed him to the king's Privy Council, and eventually made him the duke of Lennox, an extremely high station that made Stewart's Catholicism untenable. Stewart converted to Presbyterianism, which cost him his family and titles in France. He spoke of his unbreakable faithfulness to King James and longed to die for him in order to prove his devotion. When he finally did die, his embalmed heart was sent to James, per his instructions, and James wrote a poem to memorialize him.

Following Stewart's death, King James did his duty and married Anne of Denmark, establishing strong ties to Protestant Europe, and had eight children with her. Throughout their marriage, he took other lovers. Although he was also romantically linked to one other woman, there were several men with whom he maintained long-running affairs. One of them, George Villiers, seemed to be the second great love of James's life. The king named him the duke of Buckingham, making him the first commoner to receive such a high station in over a century. Recent renovations in one of James's former royal houses revealed a secret passage between the bedchambers of the king and the duke. In their private correspondence, James referred to the duke as his wife and himself as the duke's husband and spoke of a "sorrowful widow's life" should he have to live without him.[35] I bet few Protestants realize that the King James Bible they read from every Sunday is named after a man who was super gay.

Although James was clearly drawn to sexual and romantic relationships with other men that involved deep love, affection, and personal devotion, he may have also enjoyed sex with women on occasion. He fathered eight children with his wife and had at least one extramarital relationship with a woman, or so historians believe. He could have been at least somewhat bisexual, or maybe he's just one of the great many closeted gay men throughout history who were able to maintain their marriage and family life while engaging their attraction to men, usually in a clandestine manner.

While we can muse about the morality of the marital infidelity by the "divinely anointed" king, his homosexuality did not impair or inhibit his reign, even though he made little effort to hide it, at least from his inner circle. Given the rampant homophobia and state-sanctioned violence against queer persons that we have come to expect from earlier times, this seems shocking to us now. We might have thought that James would go to great lengths to conceal his lovers at the risk of losing his throne or becoming subject to blackmail and coercion. But no. James and his lovers left abundant evidence of their mutual devotion, and historians continue to unearth that evidence hundreds of years later. James's contemporaries surely knew about his private life. While he was still a young king, the Scottish nobles did try to keep him from his first lover, the duke of Lennox, but as he matured, he was able to assert his autonomy and manage his own affairs (so to speak). He was a serious monarch, committed to peace and prosperity in his realm.

However, James also actively participated in the very oppression of queer persons that made so many of them hide in the closet. He maintained strict enforcement of sodomy laws, and he even described gay sex as "among the horrible crimes which ye are bound in conscience never to forgive." He ordered judges to grant no pardons and exercise no mercy, and he wrote in his own book that it is a crime for which even monarchs ought not show mercy.

What should we make of this hypocrisy? The practice of closeted gay men persecuting their fellow homosexuals has a long history that continues to this day and that enrages the larger queer community. From Christian ministers and priests denouncing the evils of homosexuality from the pulpit while secretly enjoying its pleasures, to conservative politicians fighting hard against gay rights while maintaining covert affairs, the fear and shame of a closeted life has turned many gay men into hypocritical monsters. For men like King James, the strategic remedy for suspicion and disapproval was to maintain a staunchly homophobic public persona, but he did not appear to suffer from private shame, regret, or embarrassment.

More relevant to our exploration here is that James's homosexuality brought absolutely no impediment to his *biological fitness*. The definition of biological fitness is not just having many children but having many *successful* children and subsequent descendants. On that score, James would earn the highest possible mark. One son inherited his throne and became Charles I, a powerful monarch who had many children of his own, including two future kings, Charles II and James II, both of James's House of Stuart. James's eldest daughter became the queen of Bohemia through marriage, and her grandson would become King George I, the first British monarch of the House of Hanover. Thus, even as the monarchy was abolished and reinstated, became Catholic and then Protestant again, and transferred houses from Stuart to Hanover, the progeny of James I remained firmly in power. James's descendants are plentiful and have married into powerful families throughout Europe. Currently, every single European monarch, most of the highest ranking peers, and thousands of regular folks all around the world are among the direct descendants of gay King James.*

Like the silent crickets, the lesbian albatrosses, and basically all bonobos, the same-sex sexuality of King James did not slow the prolific spread of his genes.

* Fun fact: Although the House of Stuart hasn't held the monarchy in Great Britain since Queen Anne's death in 1707, the hereditary succession of Jacobite pretenders to the British throne persists to this day. The current pretender is Franz von Bayern, aged ninety, head of the House of Wittelsbach and duke of Bavaria, a prominent art collector and aristocrat. Were it not for the 1701 Act of Settlement, the rightful king of England, Scotland, and Northern Ireland would once again be a gay man.

Conclusion

Let me tell you what I really think

We are now at the point in this book where you expect the author to take stock of everything and then step up on their soapbox and pontificate about what it all means and what we, as a society, should do with it all. The truth is that I haven't fully worked that out for myself, let alone what I think *everyone* should do. In fact, there almost certainly isn't a one-size-fits-all answer because the real message of this book is that *the most notable feature of the human sexual experience is diversity*. And yes, sex and gender diversity is a *feature* of our species, not a bug. Let's recap what we know.

We began our journey by exploring how early life-forms evolved diploidy and gametes as a means to constantly generate new genetic diversity, and we established that the purpose of sex is to proliferate that diversity. A key event was the differentiation of the gametes into two forms—big and small—which eventually became the basis of biological sex: maleness, femaleness, and hermaphroditism. The diversity-generating nature of sex put evolution into overdrive, and life-forms became more sophisticated and multicellular. Plants, animals, and fungi evolved and diversified into the wondrous tapestry of life, from mosses to pine trees, mushrooms to yeast, and tyrannosaurs to sea cucumbers. All the while, diversity, rather than efficiency, became the ultimate path to evolutionary success on an ever more crowded planet.

After the appearance of sexes, gendered behaviors were not far behind. Animals from fish to birds to monkeys evolved creative ways to engage their masculinity and femininity that went far beyond the gamete dichotomy.

Although sex is a physical property, gender is a social one, a feature of how animals interact with one another. Although animals also eat, drink, hunt, escape, grow, heal, play, and so on, *gender* is a special collection of traits and behaviors that are closely tied with the most central process of life: reproduction. The ultimate goal of all living things is to generate more of themselves because, more than anything else, life *wants* to continue. This is why sexuality is so important to us, too.

Speaking of sexuality, for far too long, we thought that the animal sex drive was simple, heterosexual, and focused solely on fertilization. Having removed the blinders of our biases, we now know that animals can have sex with any member of their species, regardless of sex or gender, for a whole host of reasons, and without exclusivity among mated pairs, even though it unavoidably leads to jealousy and conflict. Animal sex lives are anything but simple and, as always, diversity thrives everywhere.

But the real question is, how, if it all, does this apply to the human experience?

This is where it is absolutely essential to bear in mind that human beings are genetically and *genealogically* related to all other living things on the planet, past and present, through universal common ancestry. This is not a cliché or a metaphor; it's a literal scientific fact. Neither humans, nor any life-forms on this planet, fell from the sky or were created *de novo*. We are the products of descent with modification, commonly known as evolution, and the same evolutionary forces that shaped chimpanzees, penguins, and flatworms have shaped us.

That makes a few conclusions relatively straightforward. First, the recent explosion of diverse gender identities in humans is congruent with the gender diversity we see in other animals, and as such, we can hardly call it unnatural. Second, if gender experimentation were harmful, or merely a quirk, it would not be observed across such a wide swath of animals. The tendency for social animals to play around with behaviors and even anatomy that blurs the lines between strictly masculine and feminine forms is nearly universal because it is often a path to reproductive success. Therefore, *we expect to see this in humans as well!* It would be strange if we didn't.

In my view, further proof of this is *how quickly* gender diversity has exploded. As soon as our society became more tolerant, there it was. When curmudgeons whine that, "In my day, women were women and men were men and there was none of this trans or nonbinary stuff," they are *so close* to getting it! Yes, there was far less gender diversity a generation ago, but why? Because it was harshly punished by the larger society, *making it a poor strategy for success.* In a sense, conservatives are correct when they say that gender diversity is spreading because our society is more accepting of it now. But where they are incorrect, in my view, is their belief that there is anything unnatural about it. A quick glance at other animals proves otherwise.

I fully admit, however, that this does not say anything about whether a permissive attitude toward gender diversity is good or bad. As I said at the outset, just because something is natural doesn't mean it leads to good outcomes for a society. There is nothing unnatural about a victorious silverback gorilla slaughtering the offspring of his vanquished opponent, but we probably don't want to live in a world that is permissive toward that particular behavior.*

My take from this is that humans are—surprise, surprise—just like every other animal species in the sense that we are diversity-generating machines, and especially so with gender and sexuality. We, collectively speaking, have a programmed tendency toward experimentation with masculinity, femininity, and gendered behaviors in general. That is the "is" statement. It is up to society as a whole to come up with the "ought" statements that follow. Personally, I have yet to hear a single good reason why we should do anything but celebrate that diversity. Since *diversity wins* is one of the overarching lessons of life on Earth, who are we to argue?

Importantly, human gender diversity is not limited to behaviors and cultural artifacts such as clothing and makeup. As we learned in chapter 7, anatomical and biochemical variations are common as well. There is a bimodal range for the expression of enzymes and hormone receptors, and the size and shape of our breasts, genitals, and brain areas. None of us are *perfectly* masculine or feminine for all of our features.

---

* Lucky for us, there isn't much evidence that humans have such murderous instincts the way that gorillas clearly do. However, it is true that children have a higher chance of being abused or neglected by a stepparent than a biological one. But, countering this fact is the statistic that children are *less* likely to be abused by adoptive parents than biological ones. Humans are inclined to *care* for young children, regardless of their parentage, rather than harm them.

That is why I say that sex is not a binary, probably the most controversial claim in this book, and where I lose the support of many who are otherwise in agreement with the rest. The insistence on considering sex as a pure binary stems from the true dichotomy of the gametes. There are only two kinds, big and small, and each of us makes only one of them.* Therefore, we are all either the male or the female sex, so goes the reasoning. I challenge this because it centers the gametes above every other cell and organ throughout the body, all of which are far more important to the health and daily life of the individual. For those who, for whatever reason, choose to focus solely on sperm and eggs, there is a perfectly suitable term—*gametic sex*— while those interested in sex-influenced biology throughout the body can stick with *biological sex*. It is only in the latter case that I consider sex a multidimensional spectrum. If you really love binaries, *gametic sex* is all yours. Enjoy it.

Speaking of spectra, sexuality is another concept that I think deserves a wholescale revision. I don't think the concept of *sexual orientation* is helpful. To be honest, I'm not even sure it really exists. Gay rights advocates have long claimed that gay people are "born this way," which implies a strict determinism that I don't see a lot of evidence for. Even in the relatively clean experimental study of homosexuality in rams, we saw that most rams, while having a preference, will happily copulate with their non-preferred sex as well, that is, if they enjoy copulating at all. It's clear that animals have a sex drive, and humans are no different, but that doesn't mean that this drive comes in specific flavors with strict boundaries. Instead, the study of animal behavior reveals flexibility, rather than aversions. Indeed, there would be little to gain from being averse to members of an entire sex.

This is increasingly being observed in humans as well. Both sexual fluidity and attractions along the pansexual spectrum have been the subject of study for quite a long time, beginning with Alfred Kinsey in the 1940s, so this

---

* There is an extremely rare intersex condition called ovotesticular syndrome, in which an individual has both ovarian and testicular tissue, most often caused by the fusion of two distinct zygotes into one early embryo, which then develops normally (except for the sex organs, if the two fused zygotes are not the same chromosomal sex). Such individuals are genetic chimeras. Only around five hundred of such cases have been reported and it's unclear how many of these individuals can produce both sperm and eggs.

really isn't new. Yet, the idea of fixed sexual orientation—gay or straight—has stubbornly persisted until the young adults of today began increasingly embracing labels such as bisexual, sexually fluid, pansexual, or heteroflexible. The latter—heteroflexibility—is the formal acknowledgment of an "orientation" that has long been revealed in sexuality studies: a strong and consistent sexual interest in the opposite sex but more than a fleeting attraction or curiosity about the same sex.

The fact that the younger generation is open to sexual experimentation argues that a more permissive culture allows sexual curiosity to flourish. Previously, the very harsh consequences of homosexuality placed a strong negative influence on the development of those attractions. Prepubescent young boys and girls who began to show interest in the same sex were met with shame, derision, and even violence, which forced them to bury those inclinations as best they could. The constant encouragement of heterosexual liaisons helped to nurture those nascent curiosities and their sexuality took shape accordingly. Now that these social influences are loosening their toxic grip, young people are developing a more fluid and diverse sexuality, exactly what we expect if the biological underpinnings of the sex drive are vague and flexible. If it's upsetting to you that a smaller portion of the population is strictly heterosexual, I recommend counseling to help you cope.

The issue of monogamy is perhaps the most fraught topic in this book. The question of whether or not a romantic relationship should be sexually exclusive has led to many fervent disagreements, judgmentalism, and bitter breakups. The issue divides even the ever-progressive queer community, with many preferring the freedom of open relationships and just as many wanting nothing to do with them. Many have begged psychologists, anthropologists, and biologists to answer the question "Are humans built for monogamy or not?"

Unfortunately, science does not provide a clear answer. On the one hand, cultures all over the world have settled on permanent and sexually exclusive marriage as the organizing unit of family life, which would seem to argue for some common biological basis. On the other hand, this appears to be a relatively recent development that many societies had imposed on them by cultural (and literal) invasion. On the one hand, men are 10 to 15 percent larger than women, and sexual dimorphism is associated with monogamy or sexually exclusive polygyny (while promiscuity correlates with little or no dimorphism). On the other hand, this size difference has been shrinking since at least 3 million years ago, when it was 50 percent in early species of *Homo*. This trend could indicate evolution *away* from sexual monogamy or polygyny. Our testes also give a mixed answer: the huge testes of chimpanzees speak to their promiscuity, while the tiny testes of gorillas speak to their faithfulness. Human testes are in between those extremes. Argh!

There are other supposed monogamy indicators whose significance scientists have debated. Our concealed ovulation, long birth interval, huge penises, and the uniquely human phenomenon of menopause (at least among apes) have all factored into this debate, with no clear consensus emerging. Having read all the arguments and analyzed the evidence and reasoning from both camps, my conclusion is simple: neither. Humans are not hardwired for *any* specific type of sexual-romantic relationship. Instead, we have a flexible constitution in this regard and can adapt to various social environments. Isn't that what we saw in our tour through human cultures and history in chapter 6? It's not so far-fetched. Remember the gorillas of Virunga? They seamlessly switched from the strict sexual exclusivity of the harem lifestyle to the promiscuity of large multimale/multifemale groups.

To fully appreciate the merit of this claim, you must consider that one of the truly unique aspects of our species is that we evolved to be *the ultimate generalists*. We are not adapted to any one specific habitat, diet, or way of life. This is perhaps the most remarkable thing about us and goes hand in hand with our massive brains and powerful intellects. Let me explain.

Just try to name an animal that can survive equally well in the barren desert, frozen tundra, sweltering rainforest, scattered islands, open grasslands, and luscious temperate woodlands. Only humans have managed this, but not through adaptation, through *innovation*. Humans are not *adapted* to live in the desert, but the Bedouins have thrived there for millennia. Humans are not *adapted* for a life at sea, yet the Bajou people have been seafaring for centuries. Humans are *niche constructors* because we do not adapt to our habitat, *we create* our habitat. Any other ape would quickly perish in the harsh climate of the arctic circle, but the Inuit do just fine because they fashion warm clothing and build igloos. Humans don't solve our environmental challenges with our bodies; we solve them with our minds. We solve them by adjusting our *behavior*.

And it goes beyond habitat. Consider also our diet. Most other animals have a fairly narrow range of foods that they can live on. Pandas eat only bamboo and koalas live almost exclusively on eucalyptus leaves. Compare this with the variety of human cuisines. Coastal peoples subside mostly on seafood, while the open prairies and steppes provide grain and big game. Rainforest tribes eat bugs and grubs, berries, and fleshy fruits, while the Himalayan diet is based on meat and dairy from yaks and goats. The point is that there is no single human diet. We are the ultimate omnivores and can thrive on a wide variety of foods and even stark differences in macronutrient profiles. With diet, as with habitat, humans do not wait around for mutations to cause bodily adaptations. Instead, we make our own way, eating what we find, building shelters when needed, and eking out whatever way of life that we can.

We also use our bodies in a whole host of different ways. Gorillas can sprint faster than us; Chimpanzees can climb trees better than us; Gibbons can jump higher than us; proboscis monkeys can swim better than us. But we're the only ones that can run and jump and climb and swim. We can also throw, push, pound, point, kick, point, dance, and sew. Most all other animals on the planet have found their place by specializing, but humans took the opposite approach. We evolved to do it all, which allowed us to spread all over the world, eat any food, thrive in any habitat, and figure out how to survive and thrive wherever we go. That's what those big brains are for. The ultimate generalists indeed.

And so it is with relationships. There is no one correct way for humans to live in community and form families. Cultures and societies around the world approach the matter of intimacy and family in their own way. We are not programmed to seek or tolerate just one kind of sexual relationship. Rather, we find deep emotional connection with each other in a rich variety of ways. Humans can be happy in strict sexual monogamy, free-spirited polyamory, and every permutation in between.

So, the conclusion of "what to do with all of this information" is really up to you. Humans, like every other animal, have diverse approaches to sex, gender, and sexuality, and new creative diversity is constantly being generated by the magic of sexual recombination. On top of it all, we have enormous brains for imaginative thinking and critical engagement.

So let's use them.

Acknowledgments

Like most scientists, I often get carried away with my ideas and curiosities. Therefore, whatever structure, coherence, and focus is found on these pages is due to the skill and insight of the editors and agents I have had the good fortune to work with over the five years it took to write this book. My former editor, Alex Littlefield, and my former agent, Marly Rusoff, were instrumental early on in helping me see that there was a book hiding somewhere in all my vague thoughts and lamentations. My current agents, Larry Weissman and Sascha Alper, in forcing me back to basics, helped shape what would become the outline and structure of this book. And finally, my editor, Matt Harper, who is as patient as he is wise, helped me enormously to connect the various ideas in this book into a cohesive narrative about how the intricacies of reproduction have given rise to the entire glorious world of sex and gender.

As for the ideas themselves, I must begin by acknowledging the work of Dr. Joan Roughgarden. After first reading *Evolution's Rainbow*, some fifteen years ago, I have never been the same. Both the scientific work that she describes, and the gender framework that she outlines to understand it all, are truly eye-opening and laid the foundation for this book. In fact, her work showed me the true power of a theoretical framework in helping to understand, or misunderstand, the intricacies of how life really works. It was Joan's astute description of the gender framework for understanding animal sexual behaviors that forever changed how I see the living world, and for that, I am eternally grateful. If this book has impacted your thinking in any way, you really have Joan to thank.

I also want to thank the countless friends and family members with whom I have discussed these topics over the years. I describe my writing style as a casual conversation with a friend, and so all those friends who have endured my "casual conversations" were serving as unofficial (and unpaid) editors, often against their will. There are too many of you to name, so rather than risking forgetting someone, let me just say, if I have ever engaged in a conversation with you about sex, gender, sexuality, or sexual relationships, either about animals or people, that means that I value your perspective, and you have my sincere appreciation for sharing it with me.

I must also recognize the many students that have come through my course on the biology of sex, gender, and sexuality at John Jay College. I gained so much from your unique voice, your bold ideas, and your clever writing. Even your diligent literature research impacted the content of this book in countless ways, big and small. In fact, it is difficult to imagine this book ever having been written, were it not for the input and influence of my students at John Jay College. They are among the smartest, most insightful humans I have ever met.

Notes

Introduction: The State of Affairs

1. Kaestle, Christine E. "Sexual orientation trajectories based on sexual attractions, partners, and identity: A longitudinal investigation from adolescence through young adulthood using a US representative sample." *Journal of Sex Research* (2019).

2. Watson, Ryan J., Christopher W. Wheldon, and Rebecca M. Puhl. "Evidence of diverse identities in a large national sample of sexual and gender minority adolescents." *Journal of Research on Adolescence* 30 (2020): 431–442.

3. Cover, Rob. "The proliferation of gender and sexual identities, categories and labels among young people: Emergent taxonomies." In Peter Aggleton, et al., eds. *Youth, Sexuality and Sexual Citizenship* (Routledge, 2018): 278–290.

4. Galupo, M. Paz, Shane B. Henise, and Nicholas L. Mercer. "'The labels don't work very well': Transgender individuals' conceptualizations of sexual orientation and sexual identity." *International Journal of Transgenderism* 17, no. 2 (2016): 93–104.

5. Diamond, Lisa M. "Sexual fluidity in male and females." *Current Sexual Health Reports* 8 (2016): 249–256.

6. Sassler, Sharon, and Daniel T. Lichter. "Cohabitation and marriage: Complexity and diversity in union-formation patterns." *Journal of Marriage and Family* 82, no. 1 (2020): 35–61.

7. Haupert, Mara L., Amanda N. Gesselman, Amy C. Moors, Helen E. Fisher, and Justin R. Garcia. "Prevalence of experiences with consensual nonmonogamous relationships: Findings from two national samples of single Americans." *Journal of Sex & Marital Therapy* 43, no. 5 (2017): 424–440.

8. Smock, Pamela J., and Christine R. Schwartz. "The demography of families: A review of patterns and change." *Journal of Marriage and Family* 82, no. 1 (2020): 9–34.

9. Lents, Nathan H. *Not So Different: Finding Human Nature in Animals* (Columbia University Press, 2016).

10. Johnson, Hunter R., Jessica A. Blandino, Beatriz C. Mercado, José A. Galván, William J. Higgins, and Nathan H. Lents. "The evolution of de novo human-specific microRNA genes on chromosome 21." *American Journal of Biological Anthropology* 178, no. 2 (2022): 223–243.

11. Lents, Nathan H. *Human Errors: A Panorama of Our Glitches, from Pointless Bones to Broken Genes* (Mariner Books, 2018).

Chapter 1: Evolution's Rainbow: Males, Females, and More

1. Watts, Phillip C., Kevin R. Buley, Stephanie Sanderson, Wayne Boardman, Claudio Ciofi, and Richard Gibson. "Parthenogenesis in Komodo dragons." *Nature* 444, no. 7122 (2006): 1021–1022.

2. Dudgeon, Christine L., Laura Coulton, Ren Bone, Jennifer R. Ovenden, and Severine Thomas. "Switch from sexual to parthenogenetic reproduction in a zebra shark." *Scientific Reports* 7, no. 1 (2017): 40537.

3. Lowe, Charles H., and John W. Wright. "Evolution of parthenogenetic species of Cnemidophorus (whiptail lizards) in western North America." *Journal of the Arizona Academy of Science* 4, no. 2 (1966): 81–87.

4. Barley, Anthony J., Tod W. Reeder, Adrián Nieto-Montes de Oca, Charles J. Cole, and Robert C. Thomson. "A new diploid parthenogenetic whiptail lizard from Sonora, Mexico, is the 'missing link' in the evolutionary transition to polyploidy." *American Naturalist* 198, no. 2 (2021): 295–309.

5. Newell, Peter C., Alvin Telser, and Maurice Sussman. "Alternative developmental pathways determined by environmental conditions in the cellular slime mold Dictyostelium discoideum." *Journal of Bacteriology* 100, no. 2 (1969): 763–768.

6. Welch, David B. Mark, Claudia Ricci, and Matthew Meselson. "Bdelloid rotifers: progress in understanding the success of an evolutionary scandal." In Isa Schön, et al., eds., *Lost Sex: The Evolutionary Biology of Parthenogenesis* (Springer, 2009): 259–279.

7. Mark Welch, David B., and Matthew Meselson. "Evidence for the evolution of bdelloid rotifers without sexual reproduction or genetic exchange." *Science* 288, no. 5469 (2000): 1211–1215.

8. Gladyshev, Eugene A., Matthew Meselson, and Irina R. Arkhipova. "Massive horizontal gene transfer in bdelloid rotifers." *Science* 320, no. 5880 (2008): 1210–1213.

9. Kothe, Erika. "Tetrapolar fungal mating types: sexes by the thousands." *FEMS Microbiology Reviews* 18, no. 1 (1996): 65–87.

10. Michiels, Nicolaas K., and L. J. Newman. "Sex and violence in hermaphrodites." *Nature* 391, no. 6668 (1998): 647–647.

11. Iwata, Eri, Yukiko Nagai, Mai Hyoudou, and Hideaki Sasaki. "Social environment and sex differentiation in the false clown anemonefish, Amphiprion ocellaris." *Zoological Science* 25, no. 2 (2008): 123–128.

12. Iwata, Eri, Kyohei Mikami, Jun Manbo, Keiko Moriya-Ito, and Hideaki Sasaki. "Social interaction influences blood cortisol values and brain aromatase genes in the protandrous false clown anemonefish, Amphiprion ocellaris." *Zoological Science* 29, no. 12 (2012): 849–855.

13. Prakash, Sanjeevi, Thipramalai T. Ajith Kumar, Thanumalaya Subramoniam, and J. Antonio Baeza. "Sexual system and sexual dimorphism in the shrimp Periclimenes

brevicarpalis (Schenkel, 1902)(Caridea: Palaemonidae), symbiotic with the sea anemone Stichodactyla haddoni (Saville-Kent, 1893) in the Gulf of Mannar, India." *Journal of Crustacean Biology* 37, no. 3 (2017): 332–339.

14. Miya, Masaki, and Takahisa Nemoto. "Reproduction, growth and vertical distribution of the meso- and bathypelagic fish Cyclothone atraria (Pisces: Gonostomatidae) in Sagami Bay, Central Japan." *Deep Sea Research Part A. Oceanographic Research Papers* 34, no. 9 (1987): 1565–1577.

15. Grafe, T. U., and Karl Eduard Linsenmair. "Protogynous sex change in the reed frog Hyperolius viridiflavus." *Copeia* (1989): 1024–1029.

16. Holleley, Clare E., Denis O'Meally, Stephen D. Sarre, Jennifer A. Marshall Graves, Tariq Ezaz, Kazumi Matsubara, Bhumika Azad, Xiuwen Zhang, and Arthur Georges. "Sex reversal triggers the rapid transition from genetic to temperature-dependent sex." *Nature* 523, no. 7558 (2015): 79–82.

17. Tingley, G. A., and R. M. Anderson. "Environmental sex determination and density-dependent population regulation in the entomogenous nematode Romanomermis culicivorax." *Parasitology* 92, no. 2 (1986): 431–449.

18. Davey, Andrew J. H., and Donald J. Jellyman. "Sex determination in freshwater eels and management options for manipulation of sex." *Reviews in Fish Biology and Fisheries* 15 (2005): 37–52.

19. Leet, Jessica K., Heather E. Gall, and Maria S. Sepúlveda. "A review of studies on androgen and estrogen exposure in fish early life stages: effects on gene and hormonal control of sexual differentiation." *Journal of Applied Toxicology* 31, no. 5 (2011): 379–398.

Chapter 2: Bending Gender

1. Carius, Hans Joachim, Tom J. Little, and Dieter Ebert. "Genetic variation in a host-parasite association: potential for coevolution and frequency-dependent selection." *Evolution* 55, no. 6 (2001): 1136–1145.

2. Avila, Vernon Lee. "A field study of nesting behavior of male bluegill sunfish (Lepomis macrochirus Rafinesque)." *American Midland Naturalist* (1976): 195–206.

3. Gross, Mart R. "Sneakers, satellites and parentals: polymorphic mating strategies in North American sunfishes." *Zeitschrift für Tierpsychologie* 60, no. 1 (1982): 1–26.

4. Neff, Bryan D., Peng Fu, and Mart R. Gross. "Sperm investment and alternative mating tactics in bluegill sunfish (Lepomis macrochirus)." *Behavioral Ecology* 14, no. 5 (2003): 634–641.

5. Dominey, Wallace J. "Maintenance of female mimicry as a reproductive strategy in bluegill sunfish (Lepomis macrochirus)." *Environmental Biology of Fishes* 6 (1981): 59–64.

6. Kodric-Brown, Astrid, and James H. Brown. "Truth in advertising: the kinds of traits favored by sexual selection." *American Naturalist* 124, no. 3 (1984): 309–323.

7. Bygott, J. David, Brian C. R. Bertram, and Jeannette P. Hanby. "Male lions in large coalitions gain reproductive advantages." *Nature* 282, no. 5741 (1979): 839–841.

8. Cooke, Lucy. *Bitch: a revolutionary guide to sex, evolution and the female animal* (Random House, 2022).

9. Darwin, Charles. *The Descent of Man, and Selection in Relation to Sex* (London: J. Murray, 1871).

10. Gross, Mart R. "Disruptive selection for alternative life histories in salmon." *Nature* 313, no. 5997 (1985): 47–48.

11. Gross, Mart R. "Salmon breeding behavior and life history evolution in changing environments." *Ecology* 72, no. 4 (1991): 1180–1186.

12. Lowther, James K. "Polymorphism in the white-throated sparrow, Zonotrichia albicollis (Gmelin)." *Canadian Journal of Zoology* 39, no. 3 (1961): 281–292.

13. Ficken, Robert W., Millicent S. Ficken, and Jack P. Hailman. "Differential aggression in genetically different morphs of the white-throated sparrow (Zonotrichia albicollis)." *Zeitschrift für Tierpsychologie* 46, no. 1 (1978): 43–57.

14. Lamichhaney, Sangeet, Guangyi Fan, Fredrik Widemo, Ulrika Gunnarsson, Doreen Schwochow Thalmann, Marc P. Hoeppner, Susanne Kerje, et al. "Structural genomic changes underlie alternative reproductive strategies in the ruff (Philomachus pugnax)." *Nature Genetics* 48, no. 1 (2016): 84–88.

15. Küpper, Clemens, Michael Stocks, Judith E. Risse, Natalie Dos Remedios, Lindsay L. Farrell, Susan B. McRae, Tawna C. Morgan, et al. "A supergene determines highly divergent male reproductive morphs in the ruff." *Nature Genetics* 48, no. 1 (2016): 79–83.

16. Hill, Jason, Erik D. Enbody, Huijuan Bi, Sangeet Lamichhaney, Weipan Lei, Juexin Chen, Chentao Wei, et al. "Low mutation load in a supergene underpinning alternative male mating strategies in ruff (Calidris pugnax)." *Molecular Biology and Evolution* 40, no. 12 (2023): msad224.

17. Sternalski, Audrey, François Mougeot, and Vincent Bretagnolle. "Adaptive significance of permanent female mimicry in a bird of prey." *Biology Letters* 8, no. 2 (2012): 167–170.

18. Sternalski, Audrey, and Vincent Bretagnolle. "Experimental evidence of specialised phenotypic roles in a mobbing raptor." *Behavioral Ecology and Sociobiology* 64 (2010): 1351–1361.

19. Slagsvold, Tore, and Glenn-Peter Sætre. "Evolution of plumage color in male pied flycatchers (Ficedula hypoleuca): evidence for female mimicry." *Evolution* 45, no. 4 (1991): 910–917.

20. Alatalo, Rauno V., Arae Lundberg, and Osmo Ritti. "Male polyterritoriality and imperfect female choice in the pied flycatcher, Ficedula hypoleuca." *Behavioral Ecology* 1, no. 2 (1990): 171–177.

21. Alatalo, Rauno V., Arne Lundberg, and Karin Ståhlbrandt. "Female mate choice in the pied flycatcher Ficedula hypoleuca." *Behavioral Ecology and Sociobiology* 14 (1984): 253–261.

22. Weber, A. William, and A. N. D. A. Vedder. "Population dynamics of the Virunga gorillas: 1959–1978." *Biological Conservation* 26, no. 4 (1983): 341–366.

23. Nsubuga, Anthony M., Martha M. Robbins, Christophe Boesch, and Linda Vigilant. "Patterns of paternity and group fission in wild multimale mountain gorilla groups."

American Journal of Physical Anthropology: The Official Publication of the American Association of Physical Anthropologists 135, no. 3 (2008): 263–274.

24. Robbins, Martha M., and Andrew M. Robbins. "Variation in the social organization of gorillas: Life history and socioecological perspectives." *Evolutionary Anthropology: Issues, News, and Reviews* 27, no. 5 (2018): 218–233.

25. de Waal, Frans. *Different: Gender through the Eyes of a Primatologist* (W. W. Norton & Company, 2022).

Chapter 3: That's Gay

1. Stack, S. H., L. Krannichfeld, and B. Romano. "An observation of sexual behavior between two male humpback whales." *Marine Mammal Science* (2024): e13119.

2. Stutt, Alastair D., and Michael T. Siva-Jothy. "Traumatic insemination and sexual conflict in the bed bug Cimex lectularius." *Proceedings of the National Academy of Sciences* 98, no. 10 (2001): 5683–5687.

3. Siva-Jothy, M. T. "Trauma, disease and collateral damage: conflict in cimicids." *Philosophical Transactions of the Royal Society B: Biological Sciences* 361, no. 1466 (2006): 269–275.

4. Ryne, Camilla. "Homosexual interactions in bed bugs: alarm pheromones as male recognition signals." *Animal Behaviour* 78, no. 6 (2009): 1471–1475.

5. Zuk, Marlene, John T. Rotenberry, and Robin M. Tinghitella. "Silent night: adaptive disappearance of a sexual signal in a parasitized population of field crickets." *Biology Letters* 2, no. 4 (2006): 521–524.

6. Heinen-Kay, Justa L., Ellen M. Urquhart, and Marlene Zuk. "Obligately silent males sire more offspring than singers in a rapidly evolving cricket population." *Biology Letters* 15, no. 7 (2019): 2019.0198.

7. Zhang, Xiao, Jack G. Rayner, Mark Blaxter, and Nathan W. Bailey. "Rapid parallel adaptation despite gene flow in silent crickets." *Nature Communications* 12, no. 1 (2021): 50.

8. Jukema, Joop, and Theunis Piersma. "Permanent female mimics in a lekking shorebird." *Biology Letters* 2, no. 2 (2006): 161–164.

9. Savage-Rumbaugh, E. Sue, and Beverly J. Wilkerson. "Socio-sexual behavior in Pan paniscus and Pan troglodytes: a comparative study." *Journal of Human Evolution* 7, no. 4 (1978): 327-IN6.

10. Wrangham, Richard W. "The evolution of sexuality in chimpanzees and bonobos." *Human Nature* 4 (1993): 47–79.

11. Moscovice, Liza R., Martin Surbeck, Barbara Fruth, Gottfried Hohmann, Adrian V. Jaeggi, and Tobias Deschner. "The cooperative sex: sexual interactions among female bonobos are linked to increases in oxytocin, proximity and coalitions." *Hormones and Behavior* 116 (2019): 104581.

12. Gómez, José M., A. Gónzalez-Megías, and M. Verdú. "The evolution of same-sex sexual behaviour in mammals." *Nature Communications* 14, no. 1 (2023): 5719.

13. *Ibid.*

14. Resko, John A., Anne Perkins, Charles E. Roselli, James A. Fitzgerald, Jerome V. A. Choate, and Fredrick Stormshak. "Endocrine correlates of partner preference behavior in rams." *Biology of Reproduction* 55, no. 1 (1996): 120–126.

15. Roselli, Charles E., Fred Stormshak, John N. Stellflug, and John A. Resko. "Relationship of serum testosterone concentrations to mate preferences in rams." *Biology of Reproduction* 67, no. 1 (2002): 263–268.

16. Roselli, Charles E., John A. Resko, and Fred Stormshak. "Hormonal influences on sexual partner preference in rams." *Archives of Sexual Behavior* 31 (2002): 43–49.

17. Roselli, Charles E., Kay Larkin, John A. Resko, John N. Stellflug, and Fred Stormshak. "The volume of a sexually dimorphic nucleus in the ovine medial preoptic area/anterior hypothalamus varies with sexual partner preference." *Endocrinology* 145, no. 2 (2004): 478–483.

18. Roselli, Charles E., Kay Larkin, Jessica M. Schrunk, and Fredrick Stormshak. "Sexual partner preference, hypothalamic morphology and aromatase in rams." *Physiology & Behavior* 83, no. 2 (2004): 233–245.

19. Roselli, Charles E., Radhika C. Reddy, and Katherine R. Kaufman. "The development of male-oriented behavior in rams." *Frontiers in Neuroendocrinology* 32, no. 2 (2011): 164–169.

20. LeVay, Simon. "A difference in hypothalamic structure between heterosexual and homosexual men." *Science* 253, no. 5023 (1991): 1034–1037.

21. Swaab, Dick Frans, and Eric Fliers. "A sexually dimorphic nucleus in the human brain." *Science* 228, no. 4703 (1985): 1112–1115.

22. Young, Lindsay C., Brenda J. Zaun, and Eric A. VanderWerf. "Successful same-sex pairing in Laysan albatross." *Biology Letters* 4, no. 4 (2008): 323–325.

23. *Ibid.*

24. Elie, Julie E., Nicolas Mathevon, and Clémentine Vignal. "Same-sex pair-bonds are equivalent to male–female bonds in a life-long socially monogamous songbird." *Behavioral Ecology and Sociobiology* 65 (2011): 2197–2208.

25. Richardson, Justin, and Peter Parnell. *And Tango Makes Three: With Audio Recording* (Simon and Schuster, 2015).

26. Davis, Lloyd S., Fiona M. Hunter, Robert G. Harcourt, and Sue Michelsen Heath. "Reciprocal homosexual mounting in adelie penguins Pygoscelis adeliae." *Emu-Austral Ornithology* 98, no. 2 (1998): 136–137.

27. Pincemy, Gwénaëlle, F. Stephen Dobson, and Pierre Jouventin. "Homosexual mating displays in penguins." *Ethology* 116, no. 12 (2010): 1210–1216.

28. Cezilly, Frank, and Ruedi G. Nager. "Comparative evidence for a positive association between divorce and extra-pair paternity in birds." *Proceedings of the Royal Society of London. Series B: Biological Sciences* 262, no. 1363 (1995): 7–12.

29. Heg, Dik, and Rob van Treuren. "Female–female cooperation in polygynous oystercatchers." *Nature* 391, no. 6668 (1998): 687–691.

30. Totterman, Bo, and Annette Harrison. "A long-term breeding trio of Pied Oystercatchers: Haematopus Longirostris." *Australian Field Ornithology* 24, no. 1 (2007): 7–12.

31. Quinn, Thomas W., J. Chris Davies, Fred Cooke, and Bradley N. White. "Genetic analysis of offspring of a female-female pair in the lesser snow goose (Chen c. caerulescens)." *Auk* 106, no. 2 (1989): 177–184.

32. *Ibid.*

33. Diamond, Jared M. "Goslings of gay geese." *Nature* (1989).

34. Hunt, George L. Jr., and Molly Warner Hunt. "Female-female pairing in western gulls (Larus occidentalis) in southern California." *Science* 196, no. 4297 (1977): 1466–1467.

35. Braithwaite, L. Wayne. "Ecological studies of the Black Swan III. Behaviour and social organization." *Wildlife Research* 8, no. 1 (1981): 135–146.

36. Brugger, C., and Michael Taborsky. "Male incubation and its effect on reproductive success in the Black Swan, Cygnus atratus." *Ethology* 96, no. 2 (1994): 138–146.

37. Monk, Julia D., Erin Giglio, Ambika Kamath, Max R. Lambert, and Caitlin E. McDonough. "An alternative hypothesis for the evolution of same-sex sexual behaviour in animals." *Nature Ecology & Evolution* 3, no. 12 (2019): 1622–1631.

Chapter 4: Monogamish

1. Westneat, David F. "Extra-pair fertilizations in a predominantly monogamous bird: genetic evidence." *Animal Behaviour* 35, no. 3 (1987): 877–886.

2. Birkhead, Tim R., J. Pellatt, and F. M. Hunter. "Extra-pair copulation and sperm competition in the zebra finch." *Nature* 334, no. 6177 (1988): 60–62.

3. Carter, C. Sue, and Allison M. Perkeybile. "The monogamy paradox: what do love and sex have to do with it?" *Frontiers in Ecology and Evolution* 6 (2018): 202.

4. Reichard, Ulrich H., and Christophe Boesch, eds. *Monogamy: Mating Strategies and Partnerships in Birds, Humans and Other Mammals* (Cambridge University Press, 2003).

5. Dixson, Alan F. *Mammalian Sexuality: The Act of Mating and the Evolution of Reproduction* (Cambridge University Press, 2021).

6. Burke, T., N. B. Davies, Michael William Bruford, and B. J. Hatchwell. "Parental care and mating behaviour of polyandrous dunnocks Prunella modularis related to paternity by DNA fingerprinting." *Nature* 338, no. 6212 (1989): 249–251.

7. Gill, Lisa F., Jaap van Schaik, Auguste M. P. von Bayern, and Manfred L. Gahr. "Genetic monogamy despite frequent extrapair copulations in 'strictly monogamous' wild jackdaws." *Behavioral Ecology* 31, no. 1 (2020): 247–260.

8. Shuster, Stephen M. "Sexual selection and mating systems." *Proceedings of the National Academy of Sciences* 106, supplement 1 (2009): 10009–10016.

9. Clutton-Brock, Timothy Hugh. "Review lecture: mammalian mating systems." *Proceedings of the Royal Society of London. Series B: Biological Sciences* 236, no. 1285 (1989): 339–372.

10. Lukas, Dieter, and Timothy Hugh Clutton-Brock. "The evolution of social monogamy in mammals." *Science* 341, no. 6145 (2013): 526–530.

11. Arnold, Stevan J., and David Duvall. "Animal mating systems: a synthesis based on selection theory." *American Naturalist* 143, no. 2 (1994): 317–348.

12. Getz, Lowell L., C. Sue Carter, and Leah Gavish. "The mating system of the prairie vole, Microtus ochrogaster: field and laboratory evidence for pair-bonding." *Behavioral Ecology and Sociobiology* 8 (1981): 189–194.

13. Sather, J. Henry. "Biology of the Great Plains muskrat in Nebraska." *Wildlife Monographs* 2 (1958): 1–35.

14. Bertram, Brian C. R. "The social system of lions." *Scientific American* 232, no. 5 (1975): 54–65.

15. Bertram, Brian C. R. "Kin selection in lions and in evolution." *Growing Points in Ethology* (1976): 281–301.

16. Solomon, Nancy G., and Loren D. Hayes. "Of Alloparental Behaviour in Mammals." In Gillian Bentley and Ruth Mace, eds. *Substitute Parents: Biological and Social Perspectives on Alloparenting in Human Societies* 3 (Berghahn, 2009): 13.

17. Montgomery, Tracy M., Erika L. Pendleton, and Jennifer E. Smith. "Physiological mechanisms mediating patterns of reproductive suppression and alloparental care in cooperatively breeding carnivores." *Physiology & Behavior* 193 (2018): 167–178.

18. Konrad, Christine M., Timothy R. Frasier, Hal Whitehead, and Shane Gero. "Kin selection and allocare in sperm whales." *Behavioral Ecology* 30, no. 1 (2019): 194–201.

19. Wiley, R. Haven. "Lekking in birds and mammals: behavioral and evolutionary issues." In Peter J. B. Slater, et al., eds. *Advances in the Study of Behavior*, vol. 20 (Academic Press, 1991): 201–291

20. Robertson, Kathy L., C. W. Runcorn, Julie K. Young, and Leah R. Gerber. "Spatial and temporal patterns of territory use of male California sea lions (Zalophus californianus) in the Gulf of California, Mexico." *Canadian Journal of Zoology* 86, no. 4 (2008): 237–244.

21. Gerber, Leah R., Manuela Gonzalez-Suarez, Claudia J. Hernandez-Camacho, Julie K. Young, and John L. Sabo. "The cost of male aggression and polygyny in California sea lions (Zalophus californianus)." *PLoS One* 5, no. 8 (2010): e12230.

22. Elorriaga-Verplancken, Fernando R., and Karina Acevedo-Whitehouse. "Potential intersexual altruistic behavior in California sea lions (Zalophus californianus)." *Aquatic Mammals* 43, no. 2 (2017): 208–213.

23. Flatz, Ramona, Manuela González-Suárez, Julie K. Young, Claudia J. Hernández-Camacho, Aaron J. Immel, and Leah R. Gerber. "Weak polygyny in California sea lions and the potential for alternative mating tactics." *PLoS One* 7, no. 3 (2012): e33654.

24. Carter, C. Sue, A. Courtney Devries, and Lowell L. Getz. "Physiological substrates of mammalian monogamy: the prairie vole model." *Neuroscience & Biobehavioral Reviews* 19, no. 2 (1995): 303–314.

25. Williams, Jessie R., Thomas R. Insel, Carroll R. Harbaugh, and C. Sue Carter. "Oxytocin administered centrally facilitates formation of a partner preference in female prairie voles (Microtus ochrogaster)." *Journal of Neuroendocrinology* 6, no. 3 (1994): 247–250.

26. De Dreu, Carsten K. W., Lindred L. Greer, Gerben A. Van Kleef, Shaul Shalvi, and Michel J. J. Handgraaf. "Oxytocin promotes human ethnocentrism." *Proceedings of the National Academy of Sciences* 108, no. 4 (2011): 1262–1266.

27. DeWall, C. Nathan, Omri Gillath, Sarah D. Pressman, Lora L. Black, Jennifer A. Bartz, Jackob Moskovitz, and Dean A. Stetler. "When the love hormone leads to violence: oxytocin increases intimate partner violence inclinations among high trait aggressive people." *Social Psychological and Personality Science* 5, no. 6 (2014): 691–697.

28. Wirobski, G., F. Range, F. S. Schaebs, R. Palme, T. Deschner, and S. Marshall-Pescini. "Endocrine changes related to dog domestication: Comparing urinary cortisol and oxytocin in hand-raised, pack-living dogs and wolves." *Hormones and Behavior* 128 (2021): 104901.

29. Oliva, Jessica Lee, Yen T. Wong, Jean-Loup Rault, Belinda Appleton, and Alan Lill. "The oxytocin receptor gene, an integral piece of the evolution of Canis familaris from Canis lupus." *Pet Behaviour Science* 2 (2016): 1–15.

30. Sands, Jennifer, and Scott Creel. "Social dominance, aggression and faecal glucocorticoid levels in a wild population of wolves, Canis lupus." *Animal Behaviour* 67, no. 3 (2004): 387–396.

31. Mazzini, Francesco, Simon W. Townsend, Zsófia Virányi, and Friederike Range. "Wolf howling is mediated by relationship quality rather than underlying emotional stress." *Current Biology* 23, no. 17 (2013): 1677–1680.

32. Tucker, Abigail. "What Can Rodents Tell Us about Why Humans Love?" *Smithsonian Magazine* (2014).

33. Ophir, Alexander G., Steven M. Phelps, Anna Bess Sorin, and Jerry O. Wolff. "Social but not genetic monogamy is associated with greater breeding success in prairie voles." *Animal Behaviour* 75, no. 3 (2008): 1143–1154.

34. Lim, Miranda M., Zuoxin Wang, Daniel E. Olazábal, Xianghui Ren, Ernest F. Terwilliger, and Larry J. Young. "Enhanced partner preference in a promiscuous species by manipulating the expression of a single gene." *Nature* 429, no. 6993 (2004): 754–757.

35. Dolotovskaya, Sofya, Christian Roos, and Eckhard W. Heymann. "Genetic monogamy and mate choice in a pair-living primate." *Scientific Reports* 10, no. 1 (2020): 20328.

36. Chova, Paula Escriche, Emilio Ferrer, Leana R. Goetze, Madison E. Dufek, Sara M. Freeman, and Karen L. Bales. "Neural and behavioral reactions to partners and strangers in monogamous female titi monkeys (Plecturocebus cupreus)." *Behavioural Brain Research* 443 (2023): 114334.

37. Zablocki-Thomas, Pauline B., Logan E. Savidge, Lynea R. Witczak, Emilio Ferrer, Brad A. Hobson, Abhijit J. Chaudhari, Sara M. Freeman, and Karen L. Bales. "Neural correlates and effect of jealousy on cognitive flexibility in the female titi monkey (Plecturocebus cupreus)." *Hormones and Behavior* 152 (2023): 105352.

38. Fernandez-Duque, Eduardo, Maren Huck, Sarie Van Belle, and Anthony Di Fiore. "The evolution of pair-living, sexual monogamy, and cooperative infant care: Insights from research on wild owl monkeys, titis, sakis, and tamarins." *American Journal of Physical Anthropology* 171 (2020): 118–173.

39. Fooden, Jack. "Systematic review of the Barbary macaque, Macaca sylvanus (Linnaeus, 1758)." *Fieldiana Zoology* 2007, no. 113 (2007): 1–60.

40. Small, Meredith F. "Alloparental behaviour in Barbary macaques, Macaca sylvanus." *Animal Behaviour* 39, no. 2 (1990): 297–306.

41. Paul, Andreas, Jutta Kuester, and Joachim Arnemann. "The sociobiology of male–infant interactions in Barbary macaques, Macaca sylvanus." *Animal Behaviour* 51, no. 1 (1996): 155–170.

42. Sosa, Sebastian. "The influence of gender, age, matriline and hierarchical rank on individual social position, role and interactional patterns in Macaca sylvanus at 'La Forêt des singes': A multilevel social network approach." *Frontiers in Psychology* 7 (2016): 529.

43. Widdig, Anja, Wolf Jürgen Streich, and Günter Tembrock. "Coalition formation among male Barbary macaques (Macaca sylvanus)." *American Journal of Primatology* 50, no. 1 (2000): 37–51.

44. Kuester, Jutta, and Andreas Paul. "Influence of male competition and female mate choice on male mating success in Barbary macaques (Macaca sylvanus)." *Behaviour* 120, no. 3–4 (1992): 192–216.

45. Berghänel, Andreas, Julia Ostner, Uta Schröder, and Oliver Schülke. "Social bonds predict future cooperation in male Barbary macaques, Macaca sylvanus." *Animal Behaviour* 81, no. 6 (2011): 1109–1116.

46. Small, Meredith F. "Promiscuity in Barbary macaques (Macaca sylvanus)." *American Journal of Primatology* 20, no. 4 (1990): 267–282.

47. Nadler, R. D. "Sexual behavior of orangutans." In Ronald D. Nadler, et al., eds. *The Neglected Ape* (Springer, 2013): 223.

48. Watts, David P. "Mountain gorilla reproduction and sexual behavior." *American Journal of Primatology* 24, no. 3-4 (1991): 211–225.

49. de Waal, Frans B. M. "Behavioral contrasts between bonobo and chimpanzee." In Paul Heltne and Linda A. Marquardt, eds. *Understanding Chimpanzees* (Harvard University Press, 1989): 154–175.

50. Stanford, C. B. "The social behavior of chimpanzees and bonobos: empirical evidence and shifting assumptions." *Current Anthropology* 39, no. 4 (1998), pp. 399–420.

51. Furuichi, Takeshi, and Hiroshi Ihobe. "Variation in male relationships in bonobos and chimpanzees." *Behaviour* 130, no. 3–4 (1994): 211–228.

52. Møller, Anders Pape. "Ejaculate quality, testes size and sperm competition in primates." *Journal of Human Evolution* 17, no. 5 (1988): 479–488.

53. Wittig, Roman M., Catherine Crockford, Tobias Deschner, Kevin E. Langergraber, Toni E. Ziegler, and Klaus Zuberbühler. "Food sharing is linked to urinary oxytocin levels and bonding in related and unrelated wild chimpanzees." *Proceedings of the Royal Society. Series B: Biological Sciences* 281, no. 1778 (2014): 20133096.

54. Lehmann, Julia, and Christophe Boesch. "Sexual differences in chimpanzee sociality." *International Journal of Primatology* 29 (2008): 65–81.

55. Parish, A. R. "Sex and food control in the 'uncommon chimpanzee': how bonobo females overcome a phylogenetic legacy of male dominance." *Ethology and Sociobiology* 15, no. 3 (1994): 157–179.

56. Manson, Joseph H., Susan Perry, and Amy R. Parish. "Nonconceptive sexual behavior in bonobos and capuchins." *International Journal of Primatology* 18 (1997): 767–786.

57. de Waal, Frans B. M. "Sex as an alternative to aggression in the bonobo." In Paul R. Abramson and Steven D. Pinkerton, eds. *Sexual Nature, Sexual Culture* (University of Chicago Press, 1995).

58. Wrangham, Richard W. "The evolution of sexuality in chimpanzees and bonobos." *Human Nature* 4 (1993): 47–79.

59. de Waal, Frans B. M. "Behavioral contrasts between bonobo and chimpanzee." In Heltne and Marquardt, eds. *Understanding Chimpanzees*: 154–175.

60. Moffett, Mark W. "Human identity and the evolution of societies." *Human Nature* 24 (2013): 219–267.

61. Staes, Nicky, Jeroen M. G. Stevens, Philippe Helsen, Mia Hillyer, Marisa Korody, and Marcel Eens. "Oxytocin and vasopressin receptor gene variation as a proximate base for inter- and intraspecific behavioral differences in bonobos and chimpanzees." *PLoS One* 9, no. 11 (2014): e113364.

Chapter 5: Sexual Animals

1. Berridge, Kent C., and Morten L. Kringelbach. "Pleasure systems in the brain." *Neuron* 86, no. 3 (2015): 646–664.

2. Georgiadis, Janniko R., and Morten L. Kringelbach. "The human sexual response cycle: brain imaging evidence linking sex to other pleasures." *Progress in Neurobiology* 98, no. 1 (2012): 49–81.

3. Lewis, Sara, and Adam South. "The evolution of animal nuptial gifts." In Marc Naguib, et al., eds. *Advances in the Study of Behavior* 44 (Academic Press, 2012): 53–97.

4. LeBas, Natasha R., and Leon R. Hockham. "An invasion of cheats: the evolution of worthless nuptial gifts." *Current Biology* 15, no. 1 (2005): 64–67.

5. Tryjanowski, Piotr, and Martin Hromada. "Do males of the great grey shrike, Lanius excubitor, trade food for extrapair copulations?" *Animal Behaviour* 69, no. 3 (2005): 529–533.

6. Arnqvist, Göran, Therésa M. Jones, and Mark A. Elgar. "Reversal of sex roles in nuptial feeding." *Nature* 424, no. 6947 (2003): 387.

7. Gonzalez, Andres, Carmen Rossini, Maria Eisner, and Thomas Eisner. "Sexually transmitted chemical defense in a moth (Utetheisa ornatrix)." *Proceedings of the National Academy of Sciences* 96, no. 10 (1999): 5570–5574.

8. Russell, Douglas G. D., William J. L. Sladen, and David G. Ainley. "Dr. George Murray Levick (1876–1956): unpublished notes on the sexual habits of the Adélie penguin." *Polar Record* 48, no. 4 (2012): 387–393.

9. Hunter, Fiona M., and Lloyd S. Davis. "Female Adelie penguins acquire nest material from extrapair males after engaging in extrapair copulations." *Auk* (1998): 526–528.

10. Baxter, Margaret Joan, and Linda Marie Fedigan. "Grooming and consort partner selection in a troop of Japanese monkeys (Macaca fuscata)." *Archives of Sexual Behavior* 8, no. 5 (1979): 445–458.

11. Gumert, Michael D. "Payment for sex in a macaque mating market." *Animal Behaviour* 74, no. 6 (2007): 1655–1667.

12. Takahashi, Hiroyuki, and Takeshi Furuichi. "Comparative study of grooming relationships among wild Japanese macaques in Kinkazan A troop and Yakushima M troop." *Primates* 39 (1998): 365–374.

13. Vasey, Paul L. "Sexual partner preference in female Japanese macaques." *Archives of Sexual Behavior* 31 (2002): 51–62.

14. Hockings, Kimberley J., Tatyana Humle, James R. Anderson, Dora Biro, Claudia Sousa, Gaku Ohashi, and Tetsuro Matsuzawa. "Chimpanzees share forbidden fruit." *PloS One* 2, no. 9 (2007): e886.

15. Gilby, Ian C., M. Emery Thompson, Jonathan D. Ruane, and Richard Wrangham. "No evidence of short-term exchange of meat for sex among chimpanzees." *Journal of Human Evolution* 59, no. 1 (2010): 44–53.

16. Gomes, Cristina M., and Christophe Boesch. "Wild chimpanzees exchange meat for sex on a long-term basis." *PloS One* 4, no. 4 (2009): e5116.

17. Santos, Laurie. "A monkey economy as irrational as ours." TED Global 2010 (Oxford, UK, July 2010).

18. Mason, Robert T., and David Crews. "Female mimicry in garter snakes." *Nature* 316, no. 6023 (1985): 59–60.

19. Shine, R., B. Phillips, H. Waye, M. T. M. R. LeMaster, and R. T. Mason. "Benefits of female mimicry in snakes." *Nature* 414, no. 6861 (2001): 267.

20. Parker, M. Rockwell, and Robert T. Mason. "How to make a sexy snake: estrogen activation of female sex pheromone in male red-sided garter snakes." *Journal of Experimental Biology* 215, no. 5 (2012): 723–730.

21. Haubruge, Eric, L. Arnaud, Jacques Mignon, and M. J. G. Gage. "Fertilization by proxy: rival sperm removal and translocation in a beetle." *Proceedings of the Royal Society of London. Series B: Biological Sciences* 266, no. 1424 (1999): 1183–1187.

22. Packer, Craig, and Anne E. Pusey. "Cooperation and competition within coalitions of male lions: Kin selection or game theory?" *Nature* 296, no. 5859 (1982): 740–742.

23. Packer, Craig, and Anne E. Pusey. "Divided we fall: cooperation among lions." *Scientific American* 276, no. 5 (1997): 52–59.

24. Bygott, J. David, Brian C. R. Bertram, and Jeannette P. Hanby. "Male lions in large coalitions gain reproductive advantages." *Nature* 282, no. 5741 (1979): 839–841.

25. Belkin, Aaron. *Gay Warriors: A Documentary History from the Ancient World to the Present* (New York University Press, 2002): 657–662.

26. Archie, E. A., J. Tung, M. Clark, J. Altmann, and S. C. Alberts. "Social affiliation matters: both same-sex and opposite-sex relationships predict survival in wild female baboons." *Proceedings of the Royal Society. Series B: Biological Sciences,* 281, no. 1793 (2014): 20141261.

27. Goffe, A. S., D. Zinner, and J. Fischer. "Sex and friendship in a multilevel society: behavioural patterns and associations between female and male Guinea baboons." *Behavioral Ecology and Sociobiology* 70 (2016): 323–336.

28. Roth, Lateefah, Peer Briken, and Johannes Fuss. "Masturbation in the animal kingdom." *Journal of Sex Research* 60, no. 6 (2023): 786–798.

29. Cenni, Camilla, Jessica B. A. Christie, Yanni Van der Pant, Noëlle Gunst, Paul L. Vasey, I. Nengah Wandia, and Jean-Baptiste Leca. "Do monkeys use sex toys? Evidence of stone tool-assisted masturbation in free-ranging long-tailed macaques." *Ethology* 128, no. 9 (2022): 632–646.

30. Dixson, Alan F., and Matthew J. Anderson. "Sexual behavior, reproductive physiology and sperm competition in male mammals." *Physiology & Behavior* 83, no. 2 (2004): 361–371.

31. Baker, R. Robin, and Mark A. Bellis. "Human sperm competition: Ejaculate adjustment by males and the function of masturbation." *Animal Behaviour* 46, no. 5 (1993): 861–885.

32. Pavličev, Mihaela, and Günter Wagner. "The evolutionary origin of female orgasm." *Journal of Experimental Zoology Part B: Molecular and Developmental Evolution* 326, no. 6 (2016): 326–337.

33. Waterman, Jane M. "The adaptive function of masturbation in a promiscuous African ground squirrel." *PLoS One* 5, no. 9 (2010): e13060.

Chapter 6: Family Values

1. Coontz, Stephanie. "The world historical transformation of marriage." *Journal of Marriage and the Family* (2004): 974–979.

2. Brodman, Janice Zarro. *Sex Rules!: Astonishing Sexual Practices and Gender Roles Around the World* (Mango Media Inc., 2017).

3. Hansen, Casper Worm, Peter Sandholt Jensen, and Christian Volmar Skovsgaard. "Modern gender roles and agricultural history: the Neolithic inheritance." *Journal of Economic Growth* 20 (2015): 365–404.

4. Mulder, Monique Borgerhoff, Samuel Bowles, Tom Hertz, Adrian Bell, Jan Beise, Greg Clark, Ila Fazzio, et al. "Intergenerational wealth transmission and the dynamics of inequality in small-scale societies." *Science* 326, no. 5953 (2009): 682–688.

5. Gregor, Thomas. *Anxious Pleasures: The Sexual Lives of an Amazonian People* (University of Chicago Press, 1987).

6. Politis, Gustavo. *Nukak: Ethnoarchaeology of an Amazonian People* (Routledge, 2016).

7. Tiffany, Sharon W., and Kathleen J. Adams. "Anthropology's 'fierce' Yanomami: narratives of sexual politics in the Amazon." *NWSA Journal* 6, no. 2 (1994): 169–196.

8. Everett, Daniel L. "Concentric circles of attachment among the Pirahã: A brief survey." In Hiltrud Otto, ed. *Different Faces of Attachment: Cultural Variations on a Universal Human Need* (Cambridge University Press, 2014): 169–186.

9. Jonsson, Hjorleifur. "Good attachment in the Asian highlands: questioning notions of 'loose women' and 'autonomous communities.'" *Critical Asian Studies* 52, no. 3 (2020): 403–428.

10. Herdt, Gilbert. "The Ethnography of Trobriand Sexual Culture in the 21st Century." *Anthropology Now* 5, no. 3 (2013): 134–140.

11. Senft, Gunter. "'Noble Savages' and the 'Islands of Love': Trobriand Islanders in Popular Publications." In Jürg Wassmann, ed. *Pacific Answers to Western Hegemony* (Routledge, 2020): 119–140.

12. Herdt, Gilbert. *Sambia Sexual Culture: Essays from the Field* (University of Chicago Press, 1999).

13. Diamond, Milton. "Selected cross-generational sexual behavior in traditional Hawai'i: A sexological ethnography." In Jay R. Feierman, ed. *Pedophilia: Biosocial Dimensions* (Springer, 1990): 422–444.

14. Vance, Carole S. "Anthropology rediscovers sexuality: A theoretical comment." *Culture, Society and Sexuality* (2002): 39–54.

15. Jonsson, Hjorleifur. "Ethnology and the issue of human diversity in mainland Southeast Asia." In N. J. Enfield, ed. *Dynamics of Human Diversity: The Case of Mainland Southeast Asia* (Pacific Linguistics, 2011): 109–122.

16. Ryan, Christopher, and Cacilda Jetha. *Sex at Dawn: How We Mate, Why We Stray, and What It Means for Modern Relationships* (HarperCollins, 2011).

17. Reeder, Greg. "Same-sex desire, conjugal constructs, and the tomb of Niankhkhnum and Khnumhotep." *World Archaeology* 32, no. 2 (2000): 193–208.

18. Parkinson, Richard B. *Poetry and Culture in Middle Kingdom Egypt. A Dark Side to Perfection* (Oxford University Press, 2002).

19. El-Menshawy, Sherine. "Notes on the human characteristics of ancient Egyptian kings." *Archiv orientální* 82, no. 3 (2014): 411.

20. Chapnik, Elaine. "'Women known for these acts': A review of the Jewish laws of lesbians." In Alan Slomowitz, ed. *Homosexuality, Transsexuality, Psychoanalysis and Traditional Judaism* (Routledge, 2019): 128–142.

21. Percy, William A. *Pederasty and Pedagogy in Archaic Greece* (University of Illinois Press, 1996).

22. Lear, Andrew. "Ancient pederasty: an introduction." In Thomas K, Hubbard, ed. *A Companion to Greek and Roman Sexualities* (Wiley, 2013): 102–127.

23. Scheidel, Walter. "A peculiar institution? Greco–Roman monogamy in global context." *History of the Family* 14, no. 3 (2009): 280–291.

24. Levin-Richardson, Sarah. *The Brothel of Pompeii: Sex, Class, and Gender at the Margins of Roman Society* (Cambridge University Press, 2019).

25. Fisher, Kate, and Rebecca Langlands. "The censorship myth and the secret museum." In Shelley Hales and Joanna Paul, eds. *Pompeii in the Public Imagination from Its Rediscovery to Today* (Oxford University Press, 2011): 301–315.

26. Stacey, Judith. "Unhitching the Horse from the Carriage: Love and Marriage Among the Mosuo." *Utah* Law Review (2009): 287.

27. Sanday, Peggy Reeves. *Women at the Center: Life in a Modern Matriarchy* (Cornell University Press, 2002).

28. Mattison, Siobhán M., Brooke Scelza, and Tami Blumenfield. "Paternal investment

and the positive effects of fathers among the matrilineal Mosuo of Southwest China."
American Anthropologist 116, no. 3 (2014): 591–610.

29. Goldstein, Melvyn C. "Fraternal polyandry and fertility in a high Himalayan valley in northwest Nepal." *Human Ecology* 4, no. 3 (1976): 223–233.

30. Beall, C. M., and M. C. Goldstein. "Tibetan fraternal polyandry: A test of sociobiological theory." *American Anthropologist* 83, no. 1 (1981): pp. 5–12.

31. Gould, Terry. *The Lifestyle: A Look at the Erotic Rites of Swingers* (Firefly Books, 2000).

32. Jenks, Richard J. "Swinging: A review of the literature." *Archives of Sexual Behavior* 27 (1998): 507–521.

33. Matsick, Jes L., Terri D. Conley, Ali Ziegler, Amy C. Moors, and Jennifer D. Rubin. "Love and sex: Polyamorous relationships are perceived more favourably than swinging and open relationships." *Psychology & Sexuality* 5, no. 4 (2014): 339–348.

34. Osmani, Noor Mohammad. "Misyar Marriage between Shariah texts, Realities and scholars Fatawa: An Analysis." (2010).

35. Zonouzi, Leila. "Temporary Marriage in Iran: Gender and Body Politics in Modern Iranian Film and Literature." *Journal of Middle East Women's Studies* 17, no. 1 (2021): 121–124.

36. Al-Krenawi, Alean. "Polygamy, Islam, and Marital Justice." In Alean Al-Krenawi, ed. *Psychosocial Impact of Polygamy in the Middle East* (Springer, 2014): 51–74.

37. Huda, Muhammad Chairul, Yusriyadi Yusriyadi, Mudjahirin Thohir, Miftahuddin Miftahuddin, and Muhammad Nazil Iqdami. "Nonmarital Sex Rituals on Mount Kemukus (Study of Legal Culture and Islamic Law Perspective)." *Samarah: Jurnal Hukum Keluarga dan Hukum Islam* 6, no. 1 (2022): 289–309.

38. Streitmatter, Roger. *Empty Without You: The Intimate Letters of Eleanor Roosevelt and Lorena Hickok* (Simon and Schuster, 1999).

39. Kessler, Pamela. *Undercover Washington: Where Famous Spies Lived, Worked, and Loved* (Capital Books, 2004).

Chapter 7: The Sex and Gender (Non)Binary

1. Rosser, B. R. Simon, J. Michael Oakes, Walter O. Bockting, and Michael Miner. "Capturing the social demographics of hidden sexual minorities: An internet study of the transgender population in the United States." *Sexuality Research & Social Policy* 4 (2007): 50–64.

2. Black, Dan, Gary Gates, Seth Sanders, and Lowell Taylor. "Demographics of the Gay and Lesbian Population in the United States: Evidence from Available Systematic Data Sources 1." *Queer Economics* (2013): 61–92.

3. McManus, Ian Christopher, James Moore, Matthew Freegard, and Richard Rawles. "Science in the Making: Right Hand, Left Hand. III: Estimating historical rates of left-handedness." *Laterality* 15, no. 1–2 (2010): 186–208.

4. Zucker, Kenneth J. "Epidemiology of gender dysphoria and transgender identity." *Sexual Health* 14, no. 5 (2017): 404–411.

5. Blackwood, Evelyn. "Sexuality and gender in certain Native American tribes: The case of cross-gender females." *Signs: Journal of Women in Culture and Society* 10, no. 1 (1984): 27–42.

6. Lang, Sabine. "Native American men-women, lesbians, two-spirits: Contemporary and historical perspectives." *Journal of Lesbian Studies* 20, no. 3–4 (2016): 299–323.

7. Jacobs, Sue-Ellen, Wesley Thomas, and Sabine Lang, eds. *Two-Spirit People: Native American Gender Identity, Sexuality, and Spirituality* (University of Illinois Press, 1997).

8. Mark, Joshua J. "LGBTQ+ in the Ancient World." World History Encyclopedia. Last modified June 25, 2021. https://www.worldhistory.org/article/1790/lgbtq-in-the-ancient -world/.

9. *Ibid.*

10. Merrim, Stephanie. "Catalina de Erauso: From Anomaly to Icon." In Francisco Javier Cévallos, et al., eds. *Coded Encounters: Writing, Gender, and Ethnicity in Colonial Latin America* (University of Massachusetts Press, 1994): 177–205.

11. Velasco, Sherry. *The Lieutenant Nun: Transgenderism, Lesbian Desire, and Catalina de Erauso* (University of Texas Press, 2001).

12. Fausto-Sterling, Anne. *Sexing the Body: Gender Politics and the Construction of Sexuality* (Basic Books, 2000).

13. de Lacoste-Utamsing, Christine, and Ralph L. Holloway. "Sexual dimorphism in the human corpus callosum." *Science* 216, no. 4553 (1982): 1431–1432.

14. Joel, Daphna. "Beyond the binary: Rethinking sex and the brain." *Neuroscience & Biobehavioral Reviews* 122 (2021): 165–175.

15. Joel, Daphna. "Male or female? Brains are intersex." *Frontiers in Integrative Neuroscience* 5 (2011): 57.

16. Hyde, Janet Shibley, Rebecca S. Bigler, Daphna Joel, Charlotte Chucky Tate, and Sari M. van Anders. "The future of sex and gender in psychology: Five challenges to the gender binary." *American Psychologist* 74, no. 2 (2019): 171.

17. Reis, Elizabeth. *Bodies in Doubt: An American History of Intersex* (John Hopkins University Press, 2021).

18. Horowicz, Edmund M. "Intersex children: Who are we really treating?" *Medical Law International* 17, no. 3 (2017): 183–218.

19. Carpenter, Morgan. "Intersex variations, human rights, and the international classification of diseases." *Health and Human Rights* 20, no. 2 (2018): 205.

20. Camporesi, Silvia, and Mika Hämäläinen. "A local criterion of fairness in sport: Comparing the property advantages of Caster Semenya and Eero Mäntyranta with implications for the construction of categories in sport." *Bioethics* 35, no. 3 (2021): 262–269.

21. Vann, Taylor. "Caster Semenya and the policing of competitive athletic advantage." *Connecticut Law Review* 53 (2021): 1019.

22. Batelaan, Krystal, and Gamal Abdel-Shehid. "On the Eurocentric nature of sex testing: The case of Caster Semenya." *Social Identities* 27, no. 2 (2021): 146–165.

23. Barker, Gillian. *Beyond Biofatalism: Human Nature for an Evolving World* (Columbia University Press, 2015).

24. Pinker, Steven. *The Better Angels of Our Nature: Why Violence Has Declined* (Penguin Books, 2012).

25. Buss, David. *Evolutionary Psychology: The New Science of the Mind* (Routledge, 2019).

Chapter 8: The Gay Gene

1. Jones, James H. *Alfred C. Kinsey: A Life* (W. W. Norton & Company, 2004).

2. Kinsey, Alfred C., Wardell R. Pomeroy, and Clyde E. Martin. *Sexual Behavior in the Human Male* (Saunders, 1948).

3. Kinsey, Alfred C., Wardell B. Pomeroy, Clyde E. Martin, and Paul H. Gebhard. *Sexual Behavior in the Human Female* (Saunders, 1953).

4. Poston, Dudley L., and Yu-Ting Chang. "The conceptualization and measurement of the homosexual, heterosexual, and bisexual populations in the United States." *Emerging Techniques in Applied Demography* (2015): 359–378.

5. Gonzales, Gilbert, Julia Przedworski, and Carrie Henning-Smith. "Comparison of health and health risk factors between lesbian, gay, and bisexual adults and heterosexual adults in the United States: Results from the National Health Interview Survey." *JAMA Internal Medicine* 176, no. 9 (2016): 1344–1351.

6. Zaza, Stephanie, Laura Kann, and Lisa C. Barrios. "Lesbian, gay, and bisexual adolescents: Population estimate and prevalence of health behaviors." *JAMA* 316, no. 22 (2016): 2355–2356.

7. Monto, Martin A., and Sophia Neuweiler. "The rise of bisexuality: US representative data show an increase over time in bisexual identity and persons reporting sex with both women and men." *Journal of Sex Research* (2023): 1–14.

8. Dodge, Brian, Debby Herbenick, M. Reuel Friedman, Vanessa Schick, Tsung-Chieh Fu, Wendy Bostwick, Elizabeth Bartelt, et al. "Attitudes toward bisexual men and women among a nationally representative probability sample of adults in the United States." *PloS One* 11, no. 10 (2016): e0164430.

9. Jabbour, Jeremy, Luke Holmes, David Sylva, Kevin J. Hsu, Theodore L. Semon, Alan M. Rosenthal, Adam Safron, et al. "Robust evidence for bisexual orientation among men." *Proceedings of the National Academy of Sciences* 117, no. 31 (2020): 18369–18377.

10. Slater, Eliot. "Birth order and maternal age of homosexuals." *Lancet* 279, no. 7220 (1962): 69–71.

11. Blanchard, Ray, and Anthony F. Bogaert. "Homosexuality in men and number of older brothers." *American Journal of Psychiatry* 153, no. 1 (1996): 27–31.

12. Bogaert, Anthony F. "Biological versus nonbiological older brothers and men's sexual orientation." *Proceedings of the National Academy of Sciences* 103, no. 28 (2006): 10771–10774.

13. Blanchard, Ray. "Fraternal birth order, family size, and male homosexuality: Meta-analysis of studies spanning 25 years." *Archives of Sexual Behavior* 47, no. 1 (2018): 1–15.

14. Blanchard, Ray, James M. Cantor, Anthony F. Bogaert, S. Marc Breedlove, and Lee Ellis. "Interaction of fraternal birth order and handedness in the development of male homosexuality." *Hormones and Behavior* 49, no. 3 (2006): 405–414.

15. Blanchard, Ray. "Fraternal birth order and the maternal immune hypothesis of male homosexuality." *Hormones and Behavior* 40, no. 2 (2001): 105–114.

16. Blanchard, Ray, and Philip Klassen. "HY antigen and homosexuality in men." *Journal of Theoretical Biology* 185, no. 3 (1997): 373–378.

17. Bogaert, Anthony F., Malvina N. Skorska, Chao Wang, José Gabrie, Adam J. MacNeil, Mark R. Hoffarth, Doug P. VanderLaan, Kenneth J. Zucker, and Ray Blanchard. "Male homosexuality and maternal immune responsivity to the Y-linked protein NLGN4Y." *Proceedings of the National Academy of Sciences* 115, no. 2 (2018): 302–306.

18. Ciani, Andrea Camperio, Francesca Corna, and Claudio Capiluppi. "Evidence for maternally inherited factors favouring male homosexuality and promoting female fecundity." *Proceedings of the Royal Society of London. Series B: Biological Sciences* 271, no. 1554 (2004): 2217–2221.

19. Ciani, Andrea Camperio, Francesca Iemmola, and Stan R. Blecher. "Genetic factors increase fecundity in female maternal relatives of bisexual men as in homosexuals." *Journal of Sexual Medicine* 6, no. 2 (2009): 449–455.

20. Hamer, Dean H., Stella Hu, Victoria L. Magnuson, Nan Hu, and Angela ML Pattatucci. "A linkage between DNA markers on the X chromosome and male sexual orientation." *Science* 261, no. 5119 (1993): 321–327.

21. Hu, Stella, Angela M. L. Pattatucci, Chavis Patterson, Lin Li, David W. Fulker, Stacey S. Cherny, Leonid Kruglyak, and Dean H. Hamer. "Linkage between sexual orientation and chromosome Xq28 in males but not in females." *Nature Genetics* 11, no. 3 (1995): 248–256.

22. Rice, George, Carol Anderson, Neil Risch, and George Ebers. "Male homosexuality: Absence of linkage to microsatellite markers at Xq28." *Science* 284, no. 5414 (1999): 665–667.

23. Ganna, Andrea, Karin J. H. Verweij, Michel G. Nivard, Robert Maier, Robbee Wedow, Alexander S. Busch, Abdel Abdellaoui, et al. "Large-scale GWAS reveals insights into the genetic architecture of same-sex sexual behavior." *Science* 365, no. 6456 (2019): eaat7693.

24. Manning, John T., Andrew J. G. Churchill, and Michael Peters. "The effects of sex, ethnicity, and sexual orientation on self-measured digit ratio (2D: 4D)." *Archives of Sexual Behavior* 36 (2007): 223–233.

25. Wilson, Glenn D. "Finger-length as an index of assertiveness in women." *Personality and Individual Differences* 4, no. 1 (1983): 111–112.

26. Ökten, Ayşenur, Mukaddes Kalyoncu, and Nilgün Yariş. "The ratio of second- and fourth-digit lengths and congenital adrenal hyperplasia due to 21-hydroxylase deficiency." *Early Human Development* 70, no. 1–2 (2002): 47–54.

27. Manning, John T., Peter E. Bundred, Darren J. Newton, and Brian F. Flanagan. "The second to fourth digit ratio and variation in the androgen receptor gene." *Evolution and Human Behavior* 24, no. 6 (2003): 399–405.

28. Brown, Windy M., Christopher J. Finn, and S. Marc Breedlove. "Sexual dimorphism in digit-length ratios of laboratory mice." *Anatomical Record: An Official Publication of the American Association of Anatomists* 267, no. 3 (2002): 231–234.

29. Galis, Frietson, Clara M. A. Ten Broek, Stefan Van Dongen, and Liliane C. D. Wijnaendts. "Sexual dimorphism in the prenatal digit ratio (2D: 4D)." *Archives of Sexual Behavior* 39 (2010): 57–62.

30. Jeevanandam, Saravanakumar, and Prathibha K. Muthu. "2D: 4D ratio and its implications in medicine." *Journal of Clinical and Diagnostic Research: JCDR* 10, no. 12 (2016): CM01.

31. Williams, Terrance J., Michelle E. Pepitone, Scott E. Christensen, Bradley M. Cooke, Andrew D. Huberman, Nicholas J. Breedlove, Tessa J. Breedlove, Cynthia L. Jordan, and S. Marc Breedlove. "Finger-length ratios and sexual orientation." *Nature* 404, no. 6777 (2000): 455–456.

32. Brown, Windy M., Christopher J. Finn, Bradley M. Cooke, and S. Marc Breedlove. "Differences in finger length ratios between self-identified 'butch' and 'femme' lesbians." *Archives of Sexual Behavior* 31 (2002): 123–127.

33. Kraemer, Bernd, Thomas Noll, Aba Delsignore, Gabriella Milos, Ulrich Schnyder, and Urs Hepp. "Finger length ratio (2D: 4D) and dimensions of sexual orientation." *Neuropsychobiology* 53, no. 4 (2006): 210–214.

34. Capps, Donald, and Nathan Steven Carlin. "The homosexual tendencies of King James: Should this matter to Bible readers today?" *Pastoral Psychology* 55 (2007): 667–699.

35. Bergeron, David M. *King James and Letters of Homoerotic Desire* (University of Iowa Press, 2002).

Recommended Reading

Over the many years I spent preparing to write this book, I read no fewer than fifty books and hundreds—perhaps thousands—of scientific research articles on the topics herein, many of them well outside of my own area of expertise. But there are a small handful of books that stand out as especially good choices for those who are interested in diving deeper into the biology of sex, gender, and sexuality, and their many social implications. Here I present some of those that impacted my thinking, expanded my knowledge, widened my perspective, and I think would be excellent choices for anyone looking to do the same.

The book that stands out more than any other when it comes to rethinking how animals construct sex and gender is *Evolution's Rainbow: Diversity, Gender, and Sexuality in Nature and People* by Joan Roughgarden. This book was foundational for the topic and written by a true scientific pioneer in the field. With her permission, I borrowed the title of this book for my first chapter, both because of its cleverness and as an homage to this distinguished scientist and impressive writer.

A more lighthearted but no less insightful book about animals is *Dr. Tatiana's Sex Advice to All Creation* by Olivia Judson. This book is structured like an advice column for animals who have questions about their sexual experiences. Dr. Tatiana, while giving her wise counsel, explains the intricacies of various modes of reproduction and, in so doing, highlights the incredible reproductive diversity in the animal kingdom.

I also cannot resist the urge to tout my own book, *Not So Different: Finding Human Nature in Animals*, published in 2016. You can definitely see the seeds for this book being planted in chapters 4 and 5, when I discuss sexual behaviors and social/sexual attachments in other animals. But those topics are surrounded by an expansive discussion of animal behavior more generally and the way that evolution shapes behavior in social species, which I think provides important context for the understanding of gender and sexuality. Readers with a broader interest in animal behavior, and in particular what it reveals about human behavior, may find *Not So Different* to be a good companion to this book.

Perhaps the most revolutionary and urgently important book on gendered behavior in animals is *Bitch: On the Female of the Species* by Lucy Cooke. This book is a long-overdue antidote to the pernicious influence of misogyny and patriarchy in the biological sciences and a fresh look at the diversity of female behaviors in a wide variety of animals. Although it will likely take a generation or two to undo the biases that male domination of science have inflicted on our perspectives on animal behavior, *Bitch* is a powerful start to that transition and should be read by everyone interested in animals, and especially those who plan to study them in their career.

And while I'm at it, Cat Bohannon's book, *Eve: How The Female Body Drove 200 Million Years of Human Evolution,* pairs excellently with Cooke's *Bitch,* which focuses more exclusively on humans, but with the same holistic lens of evolutionary perspective. From calling out sexist medical practices and research priorities, to celebrating feminine sexual gratification shame-free, *Eve* pushes back against centuries of societal repression of all things female. Every man, especially those in power, should read it.

For a book that closely scrutinizes gender and gendered behavior in primates as a way to understand our own construction of gender, I highly recommend *Different: Gender Through the Eyes of a Primatologist,* written by Frans de Waal, possibly the world's leading ape behaviorist. De Waal has spent a lifetime studying the behaviors of our closest relatives, the bonobos and the common chimpanzees, and every one of his books is simply dripping with wisdom and insight into why apes, including humans, behave the way that they do. De Waal's death in 2024 left an enormous void in the field of primatology, and his previous books, in particular *Chimpanzee Politics* (1983), *Our Inner Ape* (2005), *Are We Smart Enough to Know How Smart Animals Are?* (2016), and *Mama's Last Hug* (2019), have had profound impacts in our common understanding of emotion and cognition in primates, including humans.

For a book that is purely about same-sex sexual behaviors in animals, the most comprehensive volume available is *Biological Exuberance: Animal Ho-*

mosexuality and Natural Diversity, written by Bruce Bagemihl. This book synthesizes a massive amount of research and is lavishly illustrated. And in 2022, Eliot Schrefer published a children's book on the topic entitled *Queer Ducks (and Other Animals): The Natural World of Animal Sexuality*.

For those up to the challenge of tackling a comprehensive academic tome on the subject, Alan Dixson's *Mammalian Sexuality* is well worth the investment of time. This book is diligently referenced, making it easy to access the primary literature on which it is based. I find this particularly helpful when confronting theoretical frameworks that I am skeptical about, such as female mimicry and other reproductive behaviors that some consider deceptive.

The groundbreaking popular book on the diversity of human sexuality is far and away *Sex at Dawn: How We Mate, Why We Stray, and What It Means for Modern Relationships* by Christopher Ryan and Cacilda Jetha. This book, perhaps more than any other, brought the reality of human sexual diversity into the public consciousness in a way that could not be ignored. I highly recommend this book for anyone fascinated by chapter 6, as it is a much more detailed look at the myriad ways that human societies have structured sexual relationships throughout history and even in our primate ancestors. For a similar book that focuses less on the anthropological and evolutionary dissection of sexuality and is more of an entertaining tour through the diversity of modern sexual practices around the world, I recommend Janice Brodman's 2017 book, *Sex Rules!: Astonishing Sexual Practices and Gender Roles Around the World*.

For anyone who wants to learn more about the various intricacies of how consensual nonmonogamy often plays out in the modern context, I recommend *The Ethical Slut: A Practical Guide to Polyamory, Open Relationships & Other Adventures* by Dossie Easton and Janet W. Hardy. This book is a fascinating look at the way that modern couples and individuals are embracing the diverse possibilities of their sexuality. One word of advice: while I think this book is the best place to start, I would not let this be the end of your reading, as this topic is still very fraught. I would plan to consult as many perspectives as possible before planning your own consensually nonmonogamous adventure.

I also highly recommend the podcast and associated website *Savage Love*, by Dan Savage, for issues raised by real-life people and couples navigating the world of sex, love, and relationships in the modern world. A collection of the most insightful entries from his syndicated advice column have been published as the book, *Savage Love: Straight Answers from America's Most Popular Sex Columnist* and is well worth the read. For a book that is more focused on sex, love, and attraction in women, I recommend *Sexual Fluidity: Understanding Women's Love and Desire*, written by Lisa M. Diamond, one of the world's leading experts on the science of sexual attraction.

I also highly recommend *The Lifestyle* by Terry Gould. This book introduced the phenomenon of swinging to mainstream America and chronicles its fascinating history on US military bases. Along the same vein, *The Polyamorists Next Door: Inside Multiple-Partner Relationships and Families*, by Elisabeth Sheff, helps to demystify and destigmatize consensual nonmonogamy in the modern world.

On the other hand, for anyone interested in a historical or anthropological analysis of human sexual relationships—mating types, in scientific jargon—there are a number of academic books that you can explore. I enthusiastically recommend *The Origin of Family, Private Property and the State* by Friedrich Engels; *Family: A World History* by Mary Jo Maynes, Ann Waltner, and Birgitte Søland; and *Marriage, a History: How Love Conquered Marriage* by Stephanie Coontz. And lastly, Rui Diogo's magnum opus, *Meaning of Life, Human Nature, and Delusions*, devotes several chapters to the critical analysis of modern and premodern ideas about marriage, sex, gender, and sexuality, including taking a critical look at some of the books I list in this section.

For those interested in intersex diversity in humans, I recommend the novel *Middlesex*, by Jeffrey Eugenides, which won the Pulitzer Prize for fiction in 2002. Although not written by an intersex person, this book brought intersex into the mainstream consciousness and is a heartfelt and inspiring coming-of-age story of an intersex and transgender protagonist. On the other hand, Anne Fausto-Sterling's *Sexing the Body: Gender Politics and the Construction of Sexuality* is an authoritative, erudite, and at times very technical look at the human body from the nonbinary perspective. I also learned a great deal about the biology and history of intersex bodies, respectively, from *Fixing Sex: Intersex, Medical Authority, and Lived Experience* by Katrina Karkazis and *Bodies in Doubt: An American History of Intersex* by Elizabeth Reis. And finally, for a poignant, heartbreaking, but ultimately uplifting first-person narrative of the intersex experience, I urge you to read *Born Both: An Intersex Life* by Hida Viloria.

Index

NOTE: *Italic page numbers* indicate figures.

Homer, 169
Homo erectus, 7, 146, 272
Homo heidelbergensis, 2
Homo neanderthalensis, 2–3, 7, 251
Homo sapiens, 2–3, 146
homosexuality. *See also* same-sex behavior
    biological basis, 246–47, 252–62
    genes and environment, 243–46,
        252–54, 260–62, 269–70
    Kinsey and Kinsey scale, 103, 247–50,
        252, 280–81
    matrilinear connection and female
        fecundity, 257–60
    mechanistic basis, 88–91
    misconceptions about, 269–73
    as unnatural, 4, 72, 192–93, 241–42
    use of term, 72, 247
hooknose salmon, 54
horizontal gene transfer, 24
horses, 172
Human Errors (Lents), 8, 173*n*,
    260*n*
Human Evolution Blog, The, 66*n*
human sexual practices, 179–214
    "alternative lifestyles," 204–6
    ancient Egypt, 191–94
    ancient Greece, 169, 194–95
    ancient Rome, 195–97
    hunter-gatherers, 182–85
    Islam and, 206–8
    the Koechers, 211–14
    Mosuo people, 198–202
    the Roosevelts, 208–11
human sweet tooth, 109*n*
human uniqueness, 7
humpback whales, 71–72
hunger, 149, 150
hunter-gatherers, 182–85, 282–83
H-Y antigen, 256
hyenas, *65*, 65–66, 133*n*

Ideal Gas Law, 179
immune system, 175, 176, 255–56
"imperfect mate choice," 64

inbreeding, 26, 27, 81, 117, 120, 125, 129,
    134, 139
indigo buntings, 105
Indonesia, 207–8
induced ovulation, 17, 82, 173–74, 175
infanticide, 67, 68, 85, 139
infidelity, 193, 210, 274
in-group identification, 125
insects, 27, 36, 59, 74, 153, 154, 155, 165
interbreeding, 24
International Association of Athletics
    Federation (IAAF), 234–35
International Union for the Conservation of
    Nature (IUCN), 71
intersex, 9, 40, 228–38, 280*n*
    Semenya's case, 234–38
intraspecies, 75
in vitro fertilization (IVF), 270
Islam, 206–8
"is" vs. "ought," 11–12

jackdaws, 113–14
James VI and I, 273–76
Japanese macaques, 157–58, *158*
Japanese primatology, 144*n*
Johnson, Mike, 196*n*
Judaism, 193
Julius Caesar, 196
Jumonji, 38
"junk DNA," 260*n*
Jurassic Park (movie), 34

kangaroos, 172
karyotype, 16
KGB, 212–13
Khnumhotep, 191–92, 220
Kinsey, Alfred, 247–50, 252, 280–81
Kinsey scale, 103, 248, 249, 255
kittens, 119, 120–21
    play hunting, 151–52
Klinefelter syndrome, 231–32
koalas, 174, 283
Koecher, Karl and Hana, 211–14
Komodo dragons, 17

ABOUT
MARINER BOOKS

MARINER BOOKS traces its beginnings to 1832 when William Ticknor cofounded the Old Corner Bookstore in Boston, from which he would run the legendary firm Ticknor and Fields, publisher of Ralph Waldo Emerson, Harriet Beecher Stowe, Nathaniel Hawthorne, and Henry David Thoreau. Following Ticknor's death, Henry Oscar Houghton acquired Ticknor and Fields and, in 1880, formed Houghton Mifflin, which later merged with venerable Harcourt Publishing to form Houghton Mifflin Harcourt. HarperCollins purchased HMH's trade publishing business in 2021 and reestablished their storied lists and editorial team under the name Mariner Books.

Uniting the legacies of Houghton Mifflin, Harcourt Brace, and Ticknor and Fields, Mariner Books continues one of the great traditions in American bookselling. Our imprints have introduced an incomparable roster of enduring classics, including Hawthorne's *The Scarlet Letter*, Thoreau's *Walden*, Willa Cather's *O Pioneers!*, Virginia Woolf's *To the Lighthouse*, W.E.B. Du Bois's *Black Reconstruction*, J.R.R. Tolkien's *The Lord of the Rings*, Carson McCullers's *The Heart Is a Lonely Hunter*, Ann Petry's *The Narrows*, George Orwell's *Animal Farm* and *Nineteen Eighty-Four*, Rachel Carson's *Silent Spring*, Margaret Walker's *Jubilee*, Italo Calvino's *Invisible Cities*, Alice Walker's *The Color Purple*, Margaret Atwood's *The Handmaid's Tale*, Tim O'Brien's *The Things They Carried*, Philip Roth's *The Plot Against America*, Jhumpa Lahiri's *Interpreter of Maladies*, and many others. Today Mariner Books remains proudly committed to the craft of fine publishing established nearly two centuries ago at the Old Corner Bookstore.